INQUIRIES INTO CHEMISTRY

Third Edition

Michael R. Abraham
The University of Oklahoma

Michael J. Pavelich
Colorado School of Mines

Long Grove, Illinois

For information about this book, contact:
Waveland Press, Inc.
4180 IL Route 83, Suite 101
Long Grove, IL 60047-9580
(847) 634-0081
info@waveland.com
www.waveland.com

10-digit ISBN 1-57766-061-7
13-digit ISBN 978-1-57766-061-3

Printed in the United States of America

15

CONTENTS

Section One

Section Two

Section One

GUIDED INQUIRY EXPERIMENTS

A. INTRODUCTION

This section of the laboratory manual contains experiments called Guided Inquiry experiments. They cover many of the main areas of chemistry you will encounter in the lecture portion of the course (i.e., mass relationship in chemical reactions, heats of reactions, equilibria). As with any laboratory course, the experiments are designed to help you learn some of the laboratory techniques and procedures that scientists use to investigate nature. However, the experiments are designed also to serve two other purposes.

To introduce you to some of the basic concepts in chemistry.

To give you experience with some of the processes (collecting data, interpreting data, forming hypotheses and generating explanations) that a scientist uses when doing research.

Unlike some more traditional laboratory approaches, you are not expected to know anything about the subject being considered before you do the lab experiments. The purpose of the lab experiments is to introduce you to the subject area, and to give you a concrete experience with the concepts before they are discussed in a more abstract way in the lecture. Consequently many of you will be exposed to certain concepts and ideas for the first time during a laboratory activity. Use the laboratory as a learning situation where you can get your first understanding of the concepts.

Science is much more than a bunch of facts and theories. The heart of every science is research; that is, the investigation of nature. The laboratories are designed to give you experience with research. As such you will be using scientific processes such as collecting data, interpreting data, forming hypotheses, testing hypotheses and explaining results. As with any research, there are a number of explanations that will fit a set of data. You are encouraged to think independently about the data you collect. Whether you deduce the "chemically correct" explanation for your data will not be an issue. It is only important that your explanations or hypotheses follow logically from the data and that your reasoning is clearly stated.

In most experiments you will encounter a question that asks you to draw pictures to show your understanding, at the atomic/molecular level, of the phenomenon you are studying. Called Mental Models, such drawings are an important approach that most scientists use. They help us clarify our thinking and communicate it. By comparing our drawings to what others envision, we also sharpen our thinking. You are encouraged to create detailed drawings that clearly express your images of the atomic world and to share them with other students.

In these Guided Inquiry experiments you will be given specific instructions as to what experiments to conduct, and you are to answer questions about the data you collect. You should do the work in the order indicated; that is, you should answer all questions asked about one part of the experiment before you do the experimental work in the next part. Feel free to answer the questions in any way that is satisfactory to you. The "right answer" is any one that follows logically from the data and that you are comfortable with. You are encouraged to discuss your answers with your partner, other students in the lab or with the lab instructor. Such discussions are useful in that you may pick up new ideas and think more deeply about your own ideas. Your final lab report, however, should be written on your own.

After doing a particular Guided Inquiry experiment, your instructor may wish you to do the related Open Inquiry experiments found in Section Two.

This manual contains many appendices which you might find helpful. In particular, you should study the sections on Safety in the Laboratory and on Common Equipment and Procedures before doing your first experiment.

Michael R. Abraham

Michael J. Pavelich

PHYSICAL RELATIONSHIPS—Exp. B-1

Name______________________________ Lab Section __________________________

Lab Partner__________________________

Problem Statement: How are the various properties of a substance algebraically and graphically related to each other?

I. Data Collection

Measure the mass, volume (using a volumetric cylinder), length and diameter of six aluminum slugs and six brass slugs. Work in pairs to collect the data.

Sample #	Metal	Mass (g)	Volume (cm^3)	Length (cm)	Diameter (cm)

II. Data Analysis

A. Identify a relationship between two of the properties you measured for brass by graphing. Try to come up with an algebraic relationship that expresses the pattern you found. Graph paper is provided at the back of this manual.

B. On the same sheet of graph paper, graph the same variables for the aluminum. What are similarities and differences in the two graphs? Offer an explanation for this.

C. Using a different combination of variables repeat the data analysis outlined in parts II.A and II.B.

III. Interpretation

A. Using the graphs and algebraic relationships you generated in this activity, summarize the relationships you discovered. Generalize about the relationship between the graphs and their algebraic equations.

B. Mental Model—Draw a picture(s) that explains, at the level of atoms and molecules, the pattern observed in any of the relationships you investigated. Explain how your picture(s) illustrates the pattern.

UNCERTAINTY—Exp. B-2

Name________________________________ Lab Section ____________________________

Lab Partner___________________________

Problem Statements: What factors influence the variability of measurements?
How are calculated values influenced by variability in measurement?

I. Data Collection: Two Weighing Instruments

Carefully weigh the four objects indicated by your laboratory instructor using a spring scale (weigh to nearest 0.1 g) and then using a triple beam balance (weigh to nearest 0.01 g). (If you have any questions about using these instruments ask your laboratory instructor.) Record your findings in the following table.

Object	Spring Scale	Triple Beam Balance

II. Data Analysis

Compare the similarities and differences, advantages and disadvantages of each weighing instrument.

III. Data Collection: Variability in Measurement

Without comparing notes beforehand, ask five other students to weigh the four objects you weighed in part I. Have them use both the spring scale and triple beam balance. When this task is finished, share data and record them in the tables below.

Spring Scale

Object	Your Data from Part I	Student 1 Data	Student 2 Data	Student 3 Data	Student 4 Data	Student 5 Data

Triple Beam Balance

Object	Your Data from Part I	Student 1 Data	Student 2 Data	Student 3 Data	Student 4 Data	Student 5 Data

IV. Data Analysis

A. Calculate the average mass of each object as measured by the spring scale. Express how much each object varies from the average mass as a ± (plus or minus) number. (The value of ± is found by calculating how much each of the six measurements deviates from the average and then calculating the average of the deviations.) Show your calculations for one of the objects.

Object	**Mass**		
____________	________	±	________
____________	________	±	________
____________	________	±	________
____________	________	±	________

B. How would you express the precision (reproducibility) of the spring balance? What would be your reaction to someone reporting the mass of an object determined on a spring balance as 125.638 g? Explain.

C. Repeat part IV.A for the triple beam balance.

Object	**Mass**		
____________	________	±	________
____________	________	±	________
____________	________	±	________
____________	________	±	________

D. Compare how the measurements made on the spring balance and on the triple beam balance differed. List as many explanations as you can for observed differences.

V. Data Collection: Combining Measurements

Weigh each of the four objects designated by your instructor on the triple beam balance, and then weigh all four objects together. Without comparing notes ask five other students to weigh each of the four objects and combination also. When this task is finished, share data and record them in the table below.

Object	Your Data from Part I	Student 1 Data	Student 2 Data	Student 3 Data	Student 4 Data	Student 5 Data
1						
2						
3						
4						
Combination						

VI. Data Analysis

A. Calculate the maximum and minimum possible total mass for the four objects by using the individual weighings. Show your calculations below.

B. Compare the range of values (highest value—lowest value) obtained from part A with the range of values obtained for the combination weighing. Express numbers in "±" form and discuss differences and similarities.

VII. Interpretation

A. What factors influence the variability (precision) of measurements?

B. How are calculated values influenced by variability (uncertainty) in measurement?

GRAPHING RELATIONSHIPS—Exp. B-3

Name______________________________ Lab Section ___________________________

Lab Partner_________________________

Problem Statement: How are different types of relationships represented graphically and algebraically?

I. Data Collection: °F vs.°C

Arrange a ring stand with a ring, wire gauze, and a 250 mL beaker half filled with water. Measure a temperature of the water with a Celsius and a Fahrenheit thermometer (to ±0.2°). Be careful to measure the temperatures at the same time and don't let the thermometers touch the glass walls of the beaker when measuring the temperature. Measure the temperature five more times after having varied the temperature by heating or cooling the water. Record the data obtained in the table.

°F	°C

II. Data Analysis

What patterns are shown in these data? It might be helpful to graph the data. Try to come up with an algebraic equation that expresses the pattern you found.

III. Data Collection: Mass vs. Diameter

Obtain 6 rubber stoppers in different sizes from 00–13. Measure the mass of each stopper using a balance. If you are not familiar with the balance, ask your instructor to help you. Using a metric measuring stick, measure the height, top diameter, and bottom diameter of each stopper. Record your data in the table.

Stopper #	Mass (g)	Height (cm)	Top Diameter (cm)	Bottom Diameter (cm)

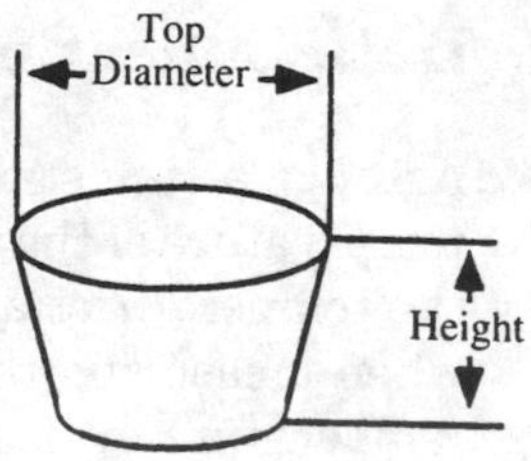

Top Diameter

Height

IV. Data Analysis

What patterns are shown in these data? It might be helpful to graph the data. Try to come up with an algebraic equation that expresses the pattern you found.

V. Data Collection: Mass vs. Volume

Obtain 6 aluminum cylinders and measure each of their masses with a balance. Record the masses in the data table. Determine the volume (to ±0.2 mL) of each aluminum cylinder by liquid displacement and record your data in the table. The liquid displacement method of determining volume is done by partially filling a 50 mL graduated cylinder with water, measuring the water, carefully submerging the aluminum in the water (tilt the cylinder and slide the aluminum down the side), and measuring the water again. The difference in the water volumes is equal to the volume of the aluminum cylinder.

Cylinder #	Mass (g)	Volume (mL)

VI. Data Analysis

What patterns are shown in these data? It might be helpful to graph the data. Try to come up with an algebraic equation that expresses the pattern you found.

VII. Interpretation

How are different types of relationships represented graphically and algebraically? Compare and contrast the similarities and differences in the graphs drawn and their algebraic representation.

HYDRATES—Exp. C-1

Name______________________________ Lab Section ____________________

Lab Partner__________________________

Problem Statements: What reaction(s) occurs when hydrated salts are heated?
How can weight data be used to confirm or deny a reaction idea?

I. Data Collection: Qualitative

A. Place a small amount (about 2cm^3) of hydrated cupric sulfate, a small amount of hydrated ferrous sulfate, and a small amount of hydrated cobalt (II) chloride into separate 150 mm Pyrex test tubes. Heat each test tube with a Bunsen burner, holding the test tubes at a slant (but not pointed at the person next to you!).

1. Describe the appearance of each hydrate before heating.

2. Describe the effects of heating on each hydrate.

B. Add a few drops of water to each cooled test tube and record your observations.

II. Data Analysis

What conclusions can be drawn from these data? (What reaction(s) do you think the compounds undergo when heated? Write chemical equations showing those ideas. How do the data support your ideas?)

III. Data Collection: Quantitative

A. Weigh a clean, dry crucible. (The proper way to prepare the crucible is to wash it, rinse with distilled water and heat for about two minutes). Add between one and four grams of hydrated cupric sulfate and record the exact weight on your data sheet. Place crucible, lid, and sample on a wire triangle on a ring stand. With the crucible lid slightly cocked, heat slowly at first, then gradually increase the heat. Do not allow the crucible to become red hot at any time. Occasionally lift the lid with tongs to observe when the dehydration is complete. When the reaction appears to be completed, continue heating for five more minutes. Let the crucible, lid, and contents cool completely, then reweigh and record the results.

B. If time permits, repeat part III.A with a different-sized sample. Collect four sets of data from other students. Make sure you have a range of different weights from 1 to 4 grams.

Hydrated Cupric Sulfate Data

Before Heating	(a)	(b)	(c)	(d)	(e)	(f)
1. mass of crucible + lid	______g	______				
2. mass of crucible + lid + sample	______	______				
3. mass of sample	______	______	______	______	______	______
After Heating						
4. mass of crucible + lid + sample	______	______				
5. mass of anhydrous salt	______	______	______	______	______	______
6. mass of water lost	______	______	______	______	______	______

IV. Data Analysis

A. What patterns are shown in these data? It might be helpful to graph the data. Try to come up with an algebraic equation that expresses the pattern you found. Explain why you chose the particular algebraic equation.

B. Mental Model—Draw pictures illustrating why the weight data show the pattern you found above. You might want to look at the chemical equation you wrote in part II and use different pictures for parts of the hydrate molecule.

C. In part II on page 19 you speculated on what reaction occurs when the hydrated copper sulfate is heated. Using this speculative chemical equation, show how you could have predicted, by calculations, what the final weights in the crucibles might be. Show calculations below for two starting weights.

V. Conclusions

Summarize the data for and against the chemical equation you wrote for the heating of hydrated copper sulfate. In particular, explain how weight data can be used as evidence for a given reaction idea. Finally, do you feel your equation reflects what nature is really doing? Briefly explain.

PRECIPITATES—Exp. C-2

Name________________________________ Lab Section ____________________________

Lab Partner____________________________

Pre-Lab Assignment: Study Appendix D.5 on filtration techniques.

Problem Statements: What is the reaction between $Co(NO_3)_2$ and Na_3PO_4?
Are all of each reactant consumed in a reaction? Why or why not?

I. Data Collection and Analysis: Qualitative

A. Dissolve a small amount (about 1 cm^3) of solid cobalt nitrate, $Co(NO_3)_2$, in about 20 mL of distilled water. Dissolve a similar amount of sodium phosphate, Na_3PO_4, in a second 20 mL of water. Describe the appearance of each solution.

B. Pour half of each solution into a third beaker and mix thoroughly. (Save the mixture and two solutions for later use.)

1. Describe the appearance of the mixture.

2. Assuming the reaction involves the coming together of dissolved ions, what are the possible identities of the solid formed in the mixture? Think up and carry out experiments which would distinguish among all of the possibilities. Describe the results of these experiments.

3. Write a chemical equation which represents the reaction and is consistent with the data obtained so far. Briefly explain your reasoning.

C. Separate the mixture by filtration (see Appendix D.5). Note the characteristics of the liquid, called the ***supernatant***, and predict all of the materials which might be dissolved in the liquid.

D. Divide the supernatant in half and test each half with the remaining $Co(NO_3)_2$ and Na_3PO_4 solutions. Describe the results.

E. What conclusions can be drawn from these data concerning the chemicals present in the supernatant? (e.g., How might changing the original amounts of $Co(NO_3)_2$ and Na_3PO_4 affect the composition of the supernatant?)

II. Data Collection: Quantitative

A. Obtain 90 mL of stock $Co(NO_3)_2$ solution in a clean, labeled beaker. Obtain 60 mL of stock Na_3PO_4 solution in a second labeled beaker. Record the concentrations in the data table below.

Clean and distilled-water rinse four 100 mL beakers and label them 1–4. Into each beaker measure 20.0 mL of stock $Co(NO_3)_2$ solution. Measure various amounts of stock Na_3PO_4 into each beaker. The suggested amounts are:

5.0, 10.0, 15.0, and 20.0 mL

Mix each solution thoroughly and allow it to stand for at least 10 minutes. Calculate the masses of salts added to each beaker and record these amounts in the table below.

B. Set up four filter funnels with filter papers and labeled collection beakers. Mix and filter each reaction mixture from part II.A through a separate funnel, collecting its supernatant in a labeled beaker. You are to collect at least 20 mL of each supernatant, clean of any precipitate. When this is done, move on to part II.C.

Data Table

Concentration of stock Na_3PO_4 solution ______________________ g/mL

Concentration of stock $Co(NO_3)_2$ solution ______________________ g/mL

Solution	mL stock Na_3PO_4	mL stock $Co(NO_3)_2$	grams Na_3PO_4	grams $Co(NO_3)_2$	Limiting Reagent by Tests	Limiting Reagent by Calculation
1						
2						
3						
4						

C. Describe the appearance of each of your supernatants. Divide each of the supernatants into two parts. Test one part with a dropperful of $Co(NO_3)_2$ and the other with a dropper full of Na_3PO_4 solution. Summarize the results of these tests below.

III. Data Analysis

A. Explain the result of this testing of the supernatants. That is, explain what these results show must have happened in each reaction mixture. (For example, why can you form more precipitate from a supernatant? Why don't all four supernatants give the same test results?) From this analysis, fill in the column labelled "Limiting Reagent by Tests" in the table on the previous page.

B. Mental Model—Draw a picture showing how the reaction can give both the precipitate and the supernatant seen in one of your reaction mixtures. In other words, illustrate your reasoning in question III.A using a picture of the reaction.

C. Predict what the limiting reactant should have been in each reaction mixture. Do this by assuming the chemical equation you wrote in part I.B.3, page 24, is correct and then calculating which reactant should run out first. Record the results of your calculations in the last column of the data table on page 25. Show the calculations for mixtures #1 and #3 below.

IV. Conclusion

Summarize the data for and against the chemical equation you wrote for the reaction between $Co(NO_3)_2$ and Na_3PO_4 in aqueous solution. In particular, explain how calculations involving weight data can be used as evidence for or against a given reaction idea. Finally, do you feel your equation reflects what nature is really doing? Briefly explain.

ZINC AND HYDROCHLORIC ACID—Exp. C-3

Name________________________________ Lab Section ____________________________

Lab Partner____________________________

Pre-Lab Assignment: Study Appendix D.1 on the use of pipets and burets.

CAUTION!!! In this experiment you will be dealing with fairly concentrated solutions of acids and bases. These substances are very corrosive, especially to the human body. Handle these substances with extreme care. If any spills occur, wash and wipe up the spill immediately. If you get any on your person, wash the affected area generously with water. See your instructor if a burning sensation is felt.

Problem Statements: What is the reaction between Zn and HCl?
How can weight and titration data be used as evidence?

I. Data Collection: Qualitative

A. Obtain three chunks of zinc (mossy or shot form) and about 100 mL of 5M hydrochloric acid solution. Put about 50 mL of distilled water in a 100 mL beaker and add one chunk of Zn. Record your observations.

Put about 50 mL of 5M HCl in a 100 mL beaker and add one chunk of Zn. Record your observations.

B. You will now run an experiment to trap the gas given off in the reaction so that it can be tested and identified. Fill a test tube with water and invert it into a 100 mL beaker of water such that no air is trapped in the test tube.

Adjust the water level in the beaker so that it just covers the mouth of the test tube. Be careful that no air enters the test tube. Add about 50 mL of 5M HCl and mix the solution. Add a fresh chunk of Zn and position the test tube such that it captures most of the gas being formed.

When the test tube is filled with gas, pull it from the solution, keeping it in a vertical position with open end down. Let the flame of a match lick over the opening of the tube. Record your observations.

II. Data Analysis

Speculate on the reaction that is occurring. Write a chemical equation to represent the reaction. Briefly explain why you chose this equation. (See Table C-2 in Appendix C.)

III. Data Collection: Quantitative

A. Clean and distilled-water rinse five 250 mL Erlenmeyer flasks. Label them 1 through 5. Obtain about 75 mL of a 5M HCl solution (The beaker used should be cleaned and rinsed with two 5 mL portions of the 5M HCl solution). Record the exact molarity of the HCl solution in the data table on page 32.

Rinse a 10 mL pipet with two 2–3 mL portions of the HCl solution (be sure to use a pipetting bulb to draw the solution into the pipet, see Appendix D.1). Pipet exactly 10.00 mL of the HCl solution into each of the five Erlenmeyer flasks. Calculate the number of moles of HCl delivered to each flask and record the values in the data table.

Using small beakers or weighing papers, weigh out two samples of ***granular*** zinc (10 to 20 mesh). Each sample should have a mass between 0.1 and 0.5 grams. Record the exact masses in the table under trials 4 and 5. Add these Zn samples slowly to Erlenmeyer flasks 4 and 5, respectively. If the reaction becomes too violent (bubbling reaching half way up the flask), stop the addition until the reaction subsides. When all the Zn has been added, cover the two flasks with watch glasses and set them aside until the reactions are complete (30 to 40 minutes). About every 10 minutes swirl the flasks to mix the reactants. In the meantime proceed with the practice titrations in parts III.B and C.

B. The purpose of the remainder of this experiment is to determine the number of moles of HCl that remain in flasks 4 and 5 after the Zn reaction is complete. You will do this by adding enough NaOH solution to just consume the remaining HCl. This analysis technique is called ***titration***. The analysis reaction is:

$$HCl\ (aq) + NaOH\ (aq) \rightarrow H_2O + NaCl\ (aq)$$

You will use the dye methyl orange to indicate when the analysis is complete. The dye changes color from pink to orange at the point where just enough NaOH has been added to consume the HCl. This point is called the ***end-point*** of the titration.

At this point # moles HCl = # moles NaOH added.

You will first titrate the HCl solutions in flasks 1 to 3 to obtain experience with recognizing the methyl orange color change and to obtain baseline data.

C. Obtain about 100 mL of stock NaOH solution (the beaker should be rinsed with two 5 mL portions of the solution to prevent contamination or dilution of the solution). Record the exact molarity of the NaOH solution in the data table. Rinse and fill a buret with the NaOH solution according to the directions in Appendix D.1. Record the initial buret reading in the table.

Add 2 to 4 drops of methyl orange indicator solution to flask 1 and swirl to mix. Slowly drain the NaOH solution from the buret into the flask, swirling the flask after each addition to mix the reactants. Continue this process until one drop of NaOH solution changes the indicator to a permanent yellow to yellow orange color. Record the final buret reading in the data table.

Calculate the total volume of NaOH solution used, the number of moles of NaOH used and the number of moles of HCl consumed by the NaOH. Record each of these in the table. Titrate the HCl solutions in flasks 2 and 3, calculate the quantities asked for, and compare the results from the three solutions for reproducibility. If the results are not within 1% of each other, talk to the instructor about ways to improve your experimental technique, or reread Appendix D.1.

D. When the reactions in flasks 4 and 5 are complete, wash down the watch glass and sides of the flask with small amounts of distilled water to return splattered material to the solution. Add 2 to 4 drops of methyl orange solution to each of the flasks and carefully titrate them with NaOH solution. (Localized white precipitate, $Zn(OH)_2$, will form where the NaOH first hits the solution. Swirl the flask until the precipitate disappears before adding the next amount of NaOH. If more than a few seconds of swirling is needed to dissolve the precipitate, you are close to the end point and the NaOH should be added a drop or two at a time.) Record the titration data in the table. Calculate the number of moles of NaOH added and number of moles of HCl titrated for each case. Show your calculations for one data set on the bottom of the next page. Collect three other sets of data on Zn reaction solutions from other students.

Data Table

Molarity of stock HCl ______________________ M

Molarity of stock NaOH ______________________ M

Trial #

	1	2	3	4	5	Other Students' Data		
# moles HCl added:								
# grams Zn added:	0.00	0.00	0.00					
Initial buret reading (mL):								
Final buret reading (mL):								
Volume of NaOH used (mL):								
# moles NaOH added:								
# moles HCl titrated:								

E. Show your calculations for one of the trials below.

IV. Data Analysis

A. What patterns are shown in the data on page 32? It might be helpful to graph the data. Find an algebraic equation that expresses this pattern. Explain why you chose this algebraic equation.

B. Offer a physical explanation for the pattern in the data. That is, discuss on a qualitative level why the amount of HCl after reaction with Zn varied as it did.

C. Mental Model—Draw a picture showing what happens to the H^+ ions in this experiment. In other words, illustrate your reasoning in question IV.B, using pictures of ions and molecules.

D. In part II. on page 30 you speculated on what balanced chemical equation best represented the reaction you observed. Show mathematically how the data on page 32 can be used to confirm or deny the chemical equation. Explain your reasoning.

V. Conclusion

Summarize the data for and against the chemical equation you wrote for the reaction of Zn with HCl. Do you feel that your equation reflects what nature is actually doing? Briefly explain.

SPECTRAL ANALYSIS FOR $Cu^{2+}(aq)$—Exp. C-4

Name________________________________ Lab Section ________________________

Lab Partner________________________

Pre-Lab Assignment: Study Appendix D.1 on burets and Appendix D.2 on the theory and use of spectrophotometers.

Problem Statements: What is the reaction between $CuSO_4$ and KOH?
How can spectroscopic data be used as evidence?

I. Data Collection and Analysis: Qualitative

A. Dissolve about 1 cm^3 of hydrated copper (II) sulfate, $CuSO_4(H_2O)_5$, in 25 mL of distilled water. Dissolve about 1 cm^3 of KOH pellets in another 25 mL of water. (**CAUTION!!!** KOH is a strong base and very corrosive to metal and flesh. Handle it with care). Describe the appearance of each solution.

B. Pour the solutions together and describe the results.

C. Assuming the reaction involves the combining of dissolved ions, what are the possible identities of the precipitate formed in the mixture? Devise and conduct experiments that will distinguish between the possibilities. Describe the results of these experiments.

D. Write a chemical equation which represents this reaction and is consistent with the data obtained so far. Briefly explain your reasoning.

II. Data Collection: Quantitative

You will be making quantitative measurements on the reaction of KOH(aq) with $CuSO_4$(aq) by using a spectrophotometer. This instrument will measure the color intensity and thus the concentrations of Cu^{2+}(aq), the ion responsible for the blue color of the solutions. Since the experiments depend on concentrations, it is important for the precision of the results that all volume and mass measurements are made with care. Review Appendix D.1 on volume measurement techniques. Also, you should be careful not to unwittingly dilute or contaminate your solutions.

A. Obtain about 120 mL of stock Cu^{2+}(aq) solution and record its concentration on page 37. This solution was prepared by dissolving a weighed amount of $CuSO_4(H_2O)_5$ to a known volume with H_2O.

B. Clean and distilled-water rinse two burets. Fill one with distilled water and drain fluid into its tip. Rinse the other with small portions of the Cu^{2+} solution, and then fill it with the Cu^{2+} solution. These burets will be used below to measure out exact volumes of water or Cu^{2+} solution.

C. Label two clean and dried 100 mL beakers. Into each weigh 0.5 to 1.8 g of KOH. Record the exact masses in the table on page 37. Add 25.00 mL of distilled water to each beaker. When the KOH has completely dissolved add 25.00 mL of the stock Cu^{2+}(aq) solution to each beaker with gentle mixing. Record the exact volumes used in the table. Let these solutions stand while you do part II.D. However, they should be stirred periodically. (To prevent contamination or loss of chemicals it is best to leave the stirring rods in their respective beakers.)

D. Prepare four different solutions having known concentrations of Cu^{2+}(aq). This is most conveniently done by diluting parts of your stock Cu^{2+}(aq) solution. The following dilutions of Cu^{2+}(aq) stock to distilled H_2O are suggested: 10.00 mL:10.00 mL, 10.00 mL:15.00 mL, 10.00 mL:20.00 mL, 10.00 mL:30.00 mL. Record the actual make-up of your diluted solutions in the table. Calculate the mole/liter concentration of each new solution and record it in the table.

E. Prepare filter paper cones for two filter funnels (see Appendix D.5.) *Do not* wet the papers with water, as this would cause a small dilution of the reaction solutions. Filter the two reaction solutions into separate, labeled, dry beakers. You are to obtain a pure sample of the supernatant (liquid part of the mixture). It may be necessary to pass the solution through the filter two or three times to obtain a sample clear of precipitate. It is only important that about 15 mL of the pure supernatant is collected. The filter paper and precipitate can be discarded.

F. You are now ready to measure the color intensity (absorbance) of blue Cu^{2+}in each of your solutions. Follow the instructions in Appendix D.2 and/or those provided by the instructor. It is suggested that the wavelength of light used for the measurements be 620 nm.

Stock $Cu^{2+}(aq)$ solution concentration ____________________ M

Reaction Solutions	Trial #1	Trial #2
mass of beaker (g)	________	________
mass of beaker + KOH	________	________
mass of KOH	________	________
mL H_2O added	________	________
mL stock Cu^{2+} added	________	________
total mL solution	________	________
absorbance at 620 nm	________	________

Standard (Diluted) $Cu^{2+}(aq)$ Solutions

	A	B	C	D
mL stock Cu^{2+}	______	______	______	______
mL H_2O added	______	______	______	______
conc. of Cu^{2+} (M)	______	______	______	______
absorbance at 620 nm	______	______	______	______

III. Data Analysis: Absorbance of Standard $Cu^{2+}(aq)$ Solutions

A. Find a pattern between the concentrations and the absorbances of your standard (diluted) $Cu^{2+}(aq)$ solutions. It may be helpful to graph the data. Find an algebraic equation that expresses this pattern, evaluate all constants in your equation and determine their units. Show and explain your work below.

B. Mental Model—Draw pictures showing what happens when light of 620 nm is passed through a Cu^{2+} solution. Show a contrast between low- and high-concentration solutions to explain the pattern you found in part III.A.

IV. Data Analysis: Reaction Solutions

A. Considering the algebraic equation and/or graph established in part III.A, determine the molar concentration of $Cu^{2+}(aq)$ in each of your reaction solutions from their absorbances. Enter these values in the table below. Make the other calculations called for in the table. Show your work for reaction Solution #1 in the space below. Complete the table by recording other students' data.

	Reaction Solution #1	Reaction Solution #2	Other Students' Data			
concentration of Cu^{2+} after reaction	________	________	________	________	________	________
moles of Cu^{2+} after reaction	________	________	________	________	________	________
moles of Cu^{2+} before reaction	________	________	________	________	________	________
moles of Cu^{2+} consumed	________	________	________	________	________	________
mass of KOH added	________	________	________	________	________	________
moles of KOH added	________	________	________	________	________	________

B. In part I.D you speculated on what reaction occurs when copper (II) sulfate reacts with potassium hydroxide. How can the data on page 37 be used to confirm or deny this reaction? Below, show calculations for your two reaction solutions. Do the data from other students fit this logic as well? Briefly explain your reasoning using numbers.

V. Conclusions

Summarize the data for and against the chemical equation you wrote for copper (II) sulfate reacting with potassium hydroxide. In particular, explain how spectroscopic data can be used as evidence for a given reaction idea. Finally, do you feel your equation reflects what nature is really doing? Briefly explain.

$CuSO_4$ AND KI—Exp. C-5

Name____________________________________ Lab Section ________________________________

Lab Partner______________________________

Pre-Lab Assignment: Study Appendix D.1 on the use of burets.

Problem Statement: What is the reaction between $CuSO_4$ and KI?

I. Data Collection and Analysis: Qualitative

A. Dissolve about 1 cm^3 of solid hydrated copper sulfate, $CuSO_4(H_2O)_5$, in 30 mL of distilled water. Dissolve about $1/2$ cm^3 of potassium iodide, KI, in a separate 30 mL of water. Describe the appearance of each solution.

B. Pour each solution into a third beaker and mix thoroughly. Describe the appearance of the mixture.

C. Filter the mixture saving both residue and supernatant. Set aside at least 15 mL of the supernatant. Wash the residue two or three times with 5 mL portions of water; these washings need not be kept. While the washing is going on, proceed with parts D and E.

D. Determine which of the ions you mixed together are the reactants in this reaction. For example: What are the four ions with which you started? Which sets of ions might have reacted? Think up and carry out experiments that will distinguish among the possibilities. Describe the results of these experiments.

E. State which ions you think are the reactants, and briefly explain your logic.

F. Describe the results of the filtration and washings done in part I.C. Do these results indicate that there are one or two product species? Briefly explain your reasoning.

G. Add several drops of the supernatant to 5 mL of cyclohexane in a large test tube. Mix vigorously and describe the results. What chemical do the results show to have been present in the supernatant? (See Table C-2 in Appendix C.)

H. Consult the Inorganic Compounds section of a CRC Handbook of Chemistry and Physics to determine the possible identities of the residue in the filter paper. List logical candidates and their published characteristics.

J. What are the products of the reaction? Briefly describe your logic.

K. Write a balanced chemical equation for the reaction, given these qualitative data.

II. Data Collection: Quantitative

A. Precisely weigh between 0.2 and 0.5 grams of $CuSO_4(H_2O)_5$ into a clean 250 mL Erlenmeyer flask. Record the exact mass in the data table on the next page. Add about 30 mL of distilled water and dissolve the copper salt.

Roughly weigh 2.5 to 3.0 grams of KI into a small beaker, add 20 mL of water to dissolve it, and then add 10 mL of 0.18 M acetic acid. Pour this mixture into the Erlenmeyer flask containing $CuSO_4$. Swirl and let stand for a few minutes to ensure that the reaction is complete.

B. Obtain about 50 mL of $Na_2S_2O_3$ solution in a clean, labeled beaker. Take precautions to not dilute or contaminate your sample of this solution. Record its exact concentration in the data table. Prepare a buret and fill it with the solution as described in Appendix D.1.

You are to titrate the $Na_2S_2O_3$ into your reaction mixture. The thiosulfate ion reacts with the I_2 in the Erlenmeyer flask according to the equation:

$$I_2 + 2\ S_2O_3^{2-} \rightarrow 2\ I^- + S_4O_6^{2-}$$

However, since iodine will react with our excess iodide as:

$$I_2 + I^- \rightarrow I_3^-$$

the actual titration reaction is:

$$I_3^- + 2\ S_2O_3^{2-} \rightarrow 3\ I^- + S_4O_6^{2-}$$

The titration end-point occurs when you have just dispelled the color of the I_2. At the end-point your solution will be a milky white. We will add starch near the end to help locate the end-point.

Titrate your reaction mixture until it turns from a dirty brown to a milky brown. Add 2 mL of starch solution, swirl to mix, and continue to titrate slowly until the blue of the starch-iodine complex is dispelled. Record the volume of thiosulfate used in the data table.

C. If time allows, titrate another reaction mixture using a different weight of Cu salt. In any case, gather enough data from other students to complete the data table. Show your calculations for one data set on the bottom of the next page.

Concentration of $Na_2S_2O_3$ = ______________________ M

Titration Data

g of $CuSO_4(H_2O)_5$					
mol of Cu^{2+}					
initial buret reading					
final buret reading					
mL of $Na_2S_2O_3$					
mol of $S_2O_3^{2-}$					
mol of I_2 (I_3^-)					

D. Show your calculations for one data set below.

III. Interpretation

A. In part I.K you wrote a balanced chemical equation for the reaction between ions in $CuSO_4$ and KI. Show mathematically how your titration data confirm or deny this equation. Finally, do you feel your equation reflects what nature is really doing in this reaction? Explain your reasoning.

B. Mental Model—Draw a picture of what is happening during this reaction. Show things like electrons transferred, bonds made, bonds broken, etc.—all the actions as you visualize them.

DISSOLUTION REACTIONS—Exp. D-1

Name________________________________ Lab Section ____________________________

Lab Partner____________________________

Problem Statement: How is heat energy related to chemical interactions?

I. Data Collection: Qualitative

A. Place about 25 mL of distilled water into a 50 mL beaker and suspend a 0.1 °C thermometer into the liquid, using a ring stand and a thermometer clamp. (Take care that the thermometer is not touching the bottom of the beaker, as any movement of the beaker could snap the thermometer.) Place a moderate amount (approximately 1 to 3 cm^3) of anhydrous magnesium sulfate, $MgSO_4$, into the water, mix vigorously for about 15 seconds and record your observations.

Repeat this procedure using fresh distilled water with each of the following compounds and record your observations.

B. Sodium Nitrate, $NaNO_3$

C. Sodium Chloride, NaCl

D. Hydrated Calcium Chloride, $CaCl_2(H_2O)_2$

E. Ammonium Nitrate, NH_4NO_3

II. Data Analysis

What conclusions can be drawn from these data? (e.g., What are the similarities and differences in the behavior of the several compounds? Does the data indicate any generalizations concerning all chemical reactions?)

III. Data Collection—Reaction: $MgSO_4(s) \xrightarrow{H_2O} Mg^{2+}(aq) + SO_4^{2-}(aq)$

A. Using a beaker, accurately weigh a 3 to 8 gram sample of $MgSO_4$ on the analytical balance. Record the exact mass in the table on page 49.

B. Suspend a 0.1 °C thermometer into a polystyrene cup so that it is about ½ inch from the bottom. Using a volumetric cylinder, add 100.0 mL of distilled water and stir while recording time versus temperature data in the table. (Data should probably be taken every 30 seconds.) After about 4 minutes add the $MgSO_4$ with vigorous mixing while continuing to record data. Mixing and the recording of data should continue for 5 to 10 minutes after the $MgSO_4$ has dissolved.

C. Determine ΔT, the temperature change for the reaction, from your data and explain how you did it (i.e., graph your data by plotting temperature versus time, draw the best curved line through the points, and explain what is occurring in each section of the curve. What part of the curve represents ΔT for the dissolution reaction?).

Data Table

	Trial #1	Trial #2
a. mass of beaker		
b. mass of beaker+$MgSO_4$		
c. mass of $MgSO_4$		

Time	Temp	Time	Temp

D. Calculate the heat, Q, of the reaction from the equation $Q=CM\ \Delta T$. Assume $C=4.18$ joules/gram °C and M is the mass of the water (take the water density as 1.00 grams/cm^3).

E. Repeat the experiment using a different sized sample of $MgSO_4$. Show your analysis of the data below. Collect four more sets of data from other students. Record all the data on the data summary table, page 51.

Summary of Data

	#1	#2	#3	#4	#5	#6
a. mass of $MgSO_4$	________	________	________	________	________	________
b. mass of water	________	________	________	________	________	________
c. ΔT	________	________	________	________	________	________
d. heat, Q	________	________	________	________	________	________

IV. Data Analysis

What pattern is shown by the data above? It might be helpful to graph the data. Give an algebraic equation that expresses this pattern.

V. Interpretation

A. How is heat energy related to chemical interactions? Use the dissolution reaction data you collected in this activity to generalize about the quantitative relationship.

B. Mental Model—Use the chemical equation: $MgSO_4(s) \xrightarrow{H_2O} Mg^{2+}(aq) + SO_4^{2-}(aq)$ to represent the dissolution reaction in part III. Use this reaction to explain how heat is released or gained by the dissolving process. Draw a picture(s) that illustrates heat gain or loss at the level of atoms and molecules. Explain how your picture(s) explains heat gain or loss.

POTASSIUM HYDROXIDE AND HYDROCHLORIC ACID
Exp. D-2

Name______________________________ Lab Section ______________________________

Lab Partner______________________________

Problem Statement: How are the heat energies for chemical reactions related to each other?

CAUTION!!! You will be working with the strong base potassium hydroxide (KOH) and the strong acid hydrochloric acid (HCl). These chemicals are extremely corrosive to metals and human flesh. If small amounts of the chemicals are spilled, immediately wipe up the spill, wash the area with water and wipe it dry. If the chemicals get on your person, wash the affected area immediately under a stream of water. If large spills occur, have the instructor handle it.

Do not place the chemicals directly on the balance pans—weigh the chemicals in beakers.

I. Data Collection—Reaction: #1: $KOH\ (s) \xrightarrow{H_2O} KOH\ (aq)$

A. Weigh between 2.5 and 5 grams of KOH pellets into a small, clean and dry beaker. Record the exact mass in the table on page 55.

Suspend a 0.1 °C thermometer into a polystyrene cup so that it is about ½ inch from the bottom. Using a volumetric cylinder, add 100.0 mL of distilled water and stir while recording time versus temperature data in the following table. (Readings should be taken every 30 seconds.) After about 3 minutes add the sample of solid KOH while continuing to stir vigorously and record data. Data should be collected at 30-second intervals for the next 5 minutes.

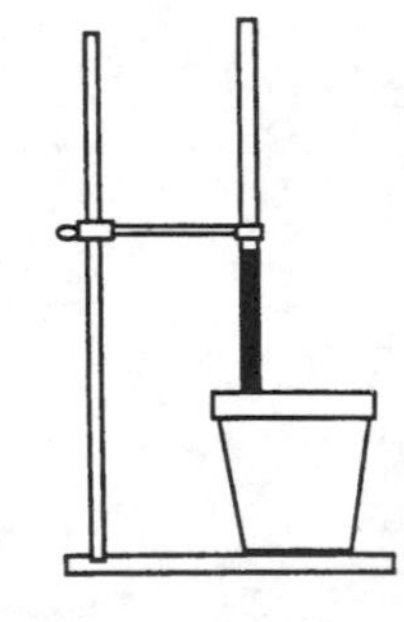

When data taking is completed, pour the KOH solution into a clean, dry labeled beaker. Cover this with a watch glass to prevent contamination. This solution will be used in part III.

B. Graph your time versus temperature data. From the graph determine the temperature change, ΔT, for the dissolving reaction. Explain how you did this. Record the ΔT value on the next page.

Reaction #1

mass of beaker (grams)________________________

mass of beaker+KOH________________________

mass of KOH ________________________

Time	Temp

ΔT for the reaction ________________________

II. Data Collection—Reaction #2: $KOH(s) + HCl(aq) \longrightarrow H_2O + KCl(aq)$

A. Weigh out another 2.5 to 5 g sample of KOH. Record the mass on the next page. Suspend the 0.1 °C thermometer as before into your clean, dry polystyrene cup.

B. Secure about 175 mL of 1.5 M HCl solution in a labeled beaker. Record the exact concentration of the solution in the table on the next page. Using a volumetric cylinder put 100.0 mL of the HCl solution into the cup. Stir and record time versus temperature data as before. After about 3 minutes add the KOH and continue to stir and record data for 5 minutes. Test the final solution with pH paper.

C. Graph these data and determine ΔT for the reaction. Was there a limiting reagent in this reaction? Briefly explain your reasoning.

III. Data Collection—Reaction #3: $KOH(aq) + HCl(aq) \longrightarrow H_2O + KCl(aq)$

A. Rinse your graduated cylinder with distilled water then measure out 50.0 mL of the KOH solution from part I into a small beaker. Rinse the cylinder and measure 50.0 mL of the 1.5 M HCl into your dry polystyrene cup. Calculate the weight and mole information called for in the table on the following page. Rinse and dry the end of the thermometer. Measure the temperature of the KOH solution and record it.

B. Rinse and dry the thermometer, then suspend it in the HCl solution. Stir and record time versus temperature data for 2 minutes. Add the 50.0 mL of KOH solution while stirring. Record time versus temperature data for about 3 minutes. Test the final solution with pH paper.

C. Graph these data and determine ΔT for the reaction. Briefly explain your reasoning. (If the initial temperatures of the two solutions are different, how should this be handled in determining ΔT for the reaction?) Was there a limiting reagent in the reaction? Briefly explain your reasoning.

Concentration of HCl solution (M) ____________________

Reaction #2 (KOH solid)

mass of beaker ____________________

mass of beaker + KOH ____________________

mass of KOH ____________________

moles HCl in 100 mL ____________________

Time	**Temp**

pH test result ____________________

ΔT for the reaction ____________________

Reaction #3 (KOH solution)

grams KOH in 50 mL ____________________

moles HCl of 50 mL ____________________

initial temp. of KOH solution ____________________

Time	**Temp**

pH test result ____________________

ΔT for the reaction ____________________

IV. Data Analysis

A. Compare the time versus temperature graphs for the three reactions. What differences exist among them? Offer a brief explanation for these differences.

B. Transfer your data to the proper table on the next page. Calculate the number of moles and heat, Q, called for. The heat is found by the equation: $Q = C\ M\ \Delta T$

As an approximation take C=4.18 joules/g °C and M as the mass of water (assume there is 1.0 g of H_2O for every 1.0 mL of solution). Show your calculations of Q below.

C. Complete the tables on the next page by collecting data from other students.

1. $KOH(s) \xrightarrow{H_2O} KOH(aq)$

a. mass of KOH(g) ______ ______ ______ ______ ______

b. moles of KOH ______ ______ ______ ______ ______

c. mL of solution ______ ______ ______ ______ ______

d. ΔT(°C) ______ ______ ______ ______ ______

e. heat, Q(J) ______ ______ ______ ______ ______

f. heat/moles KOH (J/m) ______ ______ ______ ______ ______

2. $KOH(s) + HCl(aq) \longrightarrow H_2O + KCl(aq)$

a. mass of KOH(g) ______ ______ ______ ______ ______

b. moles of KOH ______ ______ ______ ______ ______

c. moles of HCl ______ ______ ______ ______ ______

d. mL of solution ______ ______ ______ ______ ______

e. pH test ______ ______ ______ ______ ______

f. limiting reagent ______ ______ ______ ______ ______

g. ΔT(°C) ______ ______ ______ ______ ______

h. heat, Q(J) ______ ______ ______ ______ ______

i. heat/moles KOH (J/m) ______ ______ ______ ______ ______

3. $KOH(aq) + HCl(aq) \longrightarrow H_2O + KCl(aq)$

a. mass of KOH(g) ______ ______ ______ ______ ______

b. moles of KOH ______ ______ ______ ______ ______

c. moles of HCl ______ ______ ______ ______ ______

d. mL of final solution ______ ______ ______ ______ ______

e. pH test ______ ______ ______ ______ ______

f. limiting reagent ______ ______ ______ ______ ______

g. ΔT(°C) ______ ______ ______ ______ ______

h. heat, Q(J) ______ ______ ______ ______ ______

i. heat/moles KOH (J/m) ______ ______ ______ ______ ______

D. How do the data you collected on the individual reactions compare with the data collected by other students? (Identify a pattern between the amount of KOH and Q (or ΔT) in the data for each reaction individually. Find algebraic expressions for these patterns.)

V. Interpretation

A. How are the heat energies for chemical reactions related to each other? Use the three reactions studied in this experiment to support your proposed relationship (e.g., is there a mathematical relationship among the chemical equations? Is there a similar relationship among the heats?). Try to find a statement in your text that expresses the same relationship you found in the data.

B. Mental Model—Draw a picture(s) at the level of atoms and molecules that illustrates the relationship between the three reactions and their heat energies. Explain how your picture(s) illustrates the relationship.

HEATING AND COOLING BEHAVIOR—Exp. D-3

Name________________________________ Lab Section ____________________________

Lab Partner____________________________

Problem Statement: How are temperature and heat related during a phase change?

I. Data Collection: Cooling Lauric Acid

A. Obtain a sample of between 12–15 g of lauric acid in an 18 x 150 mm Pyrex test tube. Remove the stopper and place the test tube in a 400 mL beaker half-filled with tap water. Put the beaker on a ring stand and heat the water with a Bunsen burner. Do not let the water boil violently.

B. Immerse a thermometer into the sample when it becomes liquid. When the sample reaches a temperature between 65–70 °C, remove the test tube from the hot water and clamp it to a ring stand. Proceed immediately to part C.

C. Make a water bath by filling a 400 mL beaker 3/4 full of tap water at room temperature (about 25 °C).

Hold the thermometer in the center of the liquid sample in the test tube. Do not allow the thermometer to touch the bottom of the test tube. Note the temperature of the sample and the time. Immediately immerse the test tube into the water bath. Continue noting the temperature of the sample every 30 seconds while observing the state of the sample. Record your data in the table. Continue recording the temperatures until the sample is in the 30–35 °C range.

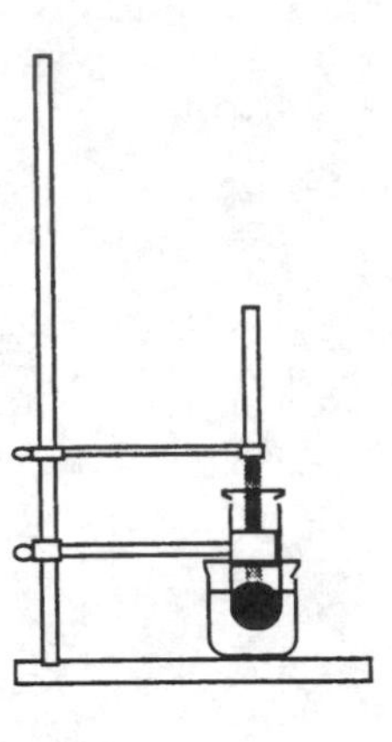

II. Data Analysis

What patterns are shown in these data? It might be helpful to graph the data.

Time	Temperature	Observations

III. Data Collection: Heating Lauric Acid

A. Allow the sample and the test tube from part I to cool to room temperature (about 25 °C). The thermometer should be embedded in the sample.

B. Make a hot water bath by filling a 400 mL beaker three-quarters of the way with water and heating it until the temperature of the water is approximately 70 °C. Turn off the burner, but leave it in position. Note the temperature of the solid sample, and the time; and then immediately immerse the test tube into the water bath. Clamp the test tube into position. Do not allow the test tube to touch the bottom of the water bath. Continue noting the temperature of the sample every 30 seconds, observing the state of the sample until the temperature of the sample is in the 60 °C range. Record your data in the table. As soon as the solid sample breaks free of the sides of the test tube, gently mix the sample with the thermometer. If the temperature of the water bath falls below 60 °C, heat it until the temperature is approximately 65 °C.

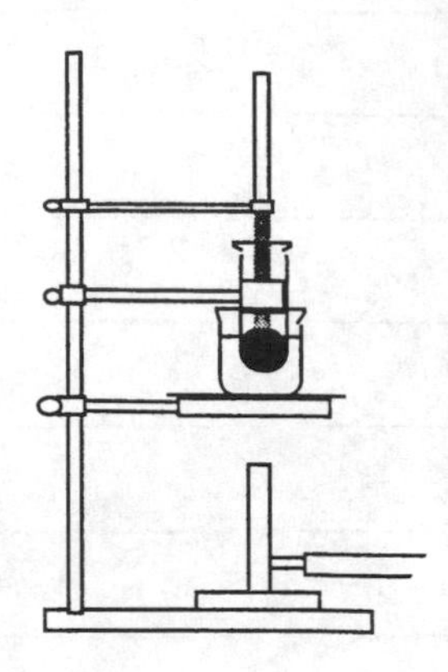

IV. Data Analysis

A. What patterns are shown in these data? It might be helpful to graph the data.

B. Compare and contrast the heating lauric acid data with the cooling lauric acid data obtained in part I. What do the data demonstrate about lauric acid?

Time	Temperature	Observations

V. Data Collection: Heating Water

Fill a 400 mL beaker half full of water at room temperature. Clamp a thermometer so that its tip is immersed about half of the depth of the water. Note the temperature of the water and the time and immediately begin heating the water with a burner. Continue noting the temperature of the water every 30 seconds while observing the state of the sample. Record your data in the table. Continue recording the temperature for about 10 minutes (or at least three minutes after boiling begins).

VI. Data Analysis

A. What patterns are shown in these data? It might be helpful to graph the data.

B. The data gathered in parts I, III, and V can be divided into zones. Sketch a theoretical time/temperature graph for water being heated from ice at -10 °C to steam at 120 °C and discuss what each zone represents.

Time	Temperature	Observations

VII. Interpretation

A. How are temperature and heat related during a phase change? Illustrate this by sketching a time/temperature graph for the heating of lauric acid from room temperature (25 °C to 150 °C. Useful information concerning lauric acid can be found in reference books (i.e., *Handbook of Chemistry and Physics*). Speculate on why the graph is shaped as it is.

B. Mental Model—Draw a picture(s) that illustrates what happens to the atoms and molecules of a solid substance when it is heated, when the solid melts, and when the melted liquid is heated. Explain your picture(s).

GAS PRESSURE AND VOLUME RELATIONSHIPS
Exp. E-1A

Name________________________________ Lab Section ____________________________

Lab Partner____________________________

Problem Statement: How are the pressure and volume of a gas sample related?

I. Data Gathering

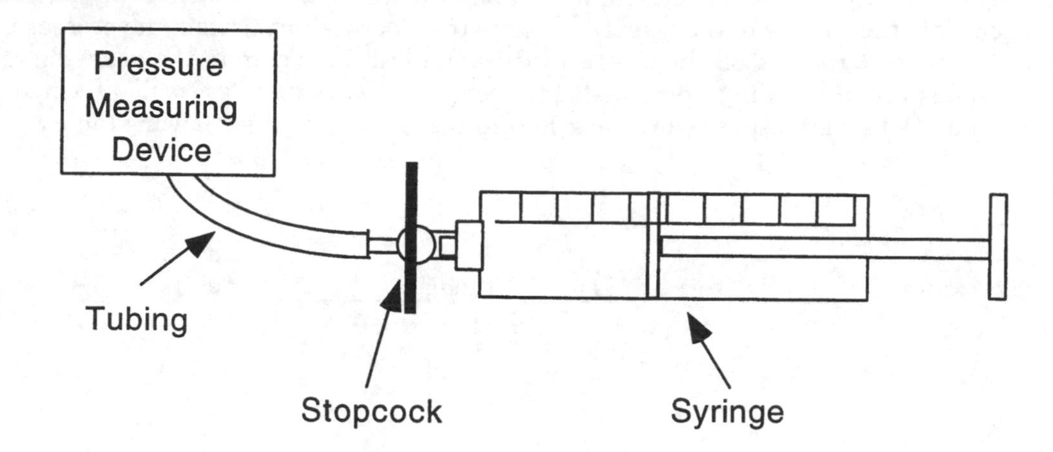

A. Obtain a pressure-measuring device as indicated by your lab instructor. Obtain a 60 mL syringe, fill it with air, and connect the syringe to the gas-measuring device as indicated in the figure. Test your apparatus for gas leaks. If you can't eliminate all leaks, see your lab instructor.

B. If necessary, calibrate your gas-measuring device as indicated by your lab instructor. Fill your syringe to the largest volume mark on the syringe and reconnect it to the gas-measuring device. What is the pressure of the trapped air in the syringe? Explain.

C. Depress the plunger of the syringe and describe the system. Is the pressure of the trapped air greater or less than atmospheric pressure? Explain.

D. Release the plunger of the syringe. Adjust the plunger to the 60 mL mark. Record the pressure reading of the trapped air in the following table. Compress the trapped air by pushing on the plunger. Note the volume of the trapped air, close the stopcock to maintain the pressure, and then note the pressure reading on the pressure device. Record these data in the following table. Take volume and pressure readings for a total of ten compressions down to as small a syringe volume as is practical. Obtain atmospheric pressure. Record these values in the following table.

Atmospheric pressure = ________ torr at ________(time), ___________(date)

Pressure Reading	Syringe Volume
____________	____________
____________	____________
____________	____________
____________	____________
____________	____________
____________	____________
____________	____________
____________	____________
____________	____________
____________	____________

II. Data Analysis

A. If necessary, calculate the total pressure of the trapped air for each reading and record it in the following table. Show how you calculate this pressure for your first reading in the space below.

B. Calculate the total volume of the trapped air for each reading and record it in the following table. Show how you calculated this volume for your first reading. (Hint: Treat the volume in the tubing and the pressure-measuring device as a cylinder, $V = \pi r^2 \times \text{length}$.)

Data Table

Pressure	Total Volume

C. What patterns are shown in these data? It might be helpful to graph the data. Try to come up with an algebraic equation that expresses the pattern you found.

III. Data Interpretation

A. How are the pressure and volume of a gas sample related?

B. Mental Model—Draw a picture(s) that explains how the pressure and volume of a gas sample are related at the level of atoms and molecules and that illustrates the observations you made in the experiment. In words, explain how your picture(s) illustrates this relationship.

GAS PRESSURE AND TEMPERATURE RELATIONSHIPS—Exp. E-1B

Name____________________________________ Lab Section ________________________________

Lab Partner______________________________

Problem Statement: How are the pressure and temperature of a gas sample related?

I. Data Gathering

A. Obtain a pressure-measuring device as indicated by your lab instructor. Assemble a 125 mL Erlenmeyer flask with thermometer, tubing, and a 1000 mL beaker as shown in the figure below. Connect this via a three-way stopcock to the pressure-measuring device and test for gas leaks. If you can't eliminate all leaks, see your lab instructor.

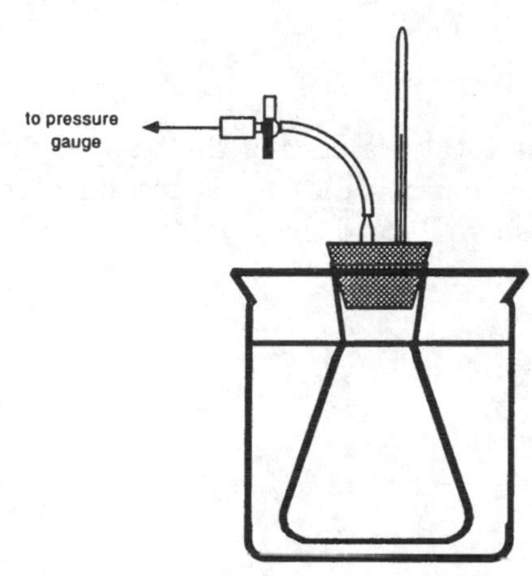

B. If necessary, calibrate your gas-measuring device as indicated by your lab instructor. Using a ring and wire gauze, support the 1000 mL beaker so that a gas burner can be used to heat the beaker. Using a clamp, suspend the flask in the beaker so that only its neck is above the beaker rim and does not touch the sides or bottom of the beaker.

Fill the beaker with tap water to about 1½ cm from the rim. Be careful not to get any water into the flask.

Adjust the stopcock so that the flask is vented to the outside. Using a gas burner, heat the water to a temperature of 80–85 °C. Use a second thermometer dipped directly into the water to monitor water temperature. The water should be stirred constantly during the heating process. The thermometer can be used to stir, but care must be taken not to break the fragile thermometer.

When the temperature reaches 80–85 °C, remove the heat. When the thermometer measuring the air temperature inside the flask reaches a maximum, adjust the stopcock to connect the flask to the pressure-measuring device while otherwise sealing off the apparatus. While continuing to gently stir the water, allow the temperature to drop about 5 °C.

C. Record the temperature and pressure reading in the following table. Allow the gas temperature to cool approximately another 5 °C while continuing to stir the water. Record the temperature of the gas sample and the pressure reading in the following table. Using the procedure outlined above, continue to record readings at approximately 5 °C intervals until a temperature of about 15 °C is reached.

NOTE:

1. If the system leaks at any time, the experiment must be restarted.
2. Cooling can be hastened by adding small amounts of ice to the water. However, to insure the temperature of the gas sample has been equalized, stir for at least 3 minutes after the ice has disappeared before taking readings.
3. Excess water can be removed from the beaker. However, the water level should be at most 3 cm from the rim.

II. Data Analysis

A. If necessary, calculate the total pressure of the trapped air for each reading and record it in the following table. Show how you calculated this pressure for your first reading in the space below. Obtain atmospheric pressure. Record these values in the following table.

Atmospheric pressure = ________ torr at ________(time), ________(date)

Pressure Reading	Temperature (°C)	Pressure of Trapped Gas
________	________	________
________	________	________
________	________	________
________	________	________
________	________	________
________	________	________
________	________	________
________	________	________
________	________	________
________	________	________

B. What patterns are shown in these data? It might be helpful to graph the data. Try to come up with an algebraic equation that expresses the pattern you found.

C. (Optional) Estimate the temperature of a gas when the pressure is reduced to zero. Discuss the significance of this temperature.

III. Data Interpretation

A. How are the pressure and temperature of a gas sample related?

B. Mental Model—Draw a picture(s) that explains how the pressure and temperature of a gas sample are related at the level of atoms and molecules and that illustrates the observations you made in the experiment. In words, explain how your picture(s) illustrates this relationship.

GAS VOLUME AND TEMPERATURE RELATIONSHIPS
Exp. E-1C

Name______________________________________ Lab Section ________________________________

Lab Partner______________________________

Problem Statement: How are the volume and temperature of a gas sample related?

I. Data Gathering

A. Using a ring and wire gauze, support a 1000 mL beaker so that a gas burner can be used to heat the beaker. Obtain a graduated "J"-tube filled with oil such that an air sample is trapped in its closed end. Suspend the J-tube with a thermometer clamp in the 1000 mL beaker filled with tap water so that the air trapped in the short end of the tube is well below the surface of the water (see figure below). Suspend a thermometer with another thermometer clamp in the water so that it is next to the air trapped in the short end. Make sure that the J-tube is arranged so that its graduations can be easily read.

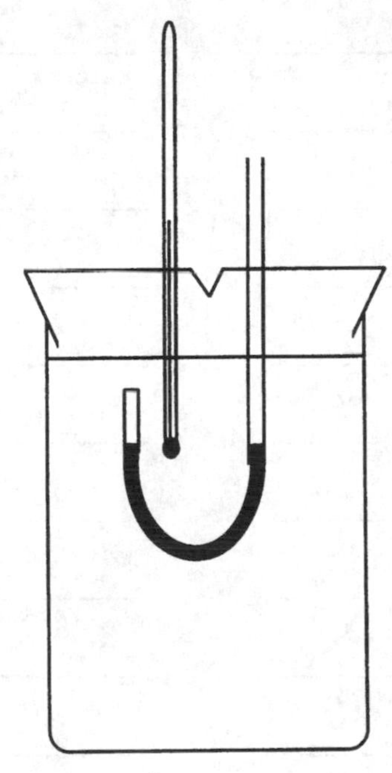

B. Read the temperature of the water and the volume of the trapped air in the J-tube. Show how you calculate the volume of the trapped air from reading the graduations from the J-tube. Record your temperature and volume readings in the following table.

C. Using a gas burner, heat the water to a temperature of 75–80 °C. The water should be constantly stirred during the heating process. When the temperature reaches 75–80 °C, remove the heat. Allow the temperature of the trapped air to equalize with the water temperature by continuing to gently stir the water and allowing the temperature to drop about 5 °C. At that point read the temperature of the water and volume of the trapped air and record these values in the following table. Using the procedure outlined above, continue to record readings at approximately 5 °C intervals until a temperature of about 15 °C is reached.

NOTE:

1. Cooling can be hastened by adding small amounts of ice to the water. However, to insure the temperature of the trapped air has been equalized, stir for at least 3 minutes after the ice has disappeared before taking readings.
2. Excess water can be removed from the beaker. However, make sure the J-tube remains submerged well below the water level.

Temperature	Volume

II. Data Analysis

A. What patterns are shown in these data? It might be helpful to graph the data. Try to come up with an algebraic equation that expresses the pattern you found.

B. (Optional) Estimate the temperature of a gas when the volume is reduced to zero. Discuss the significance of this temperature.

III. Data Interpretation

A. How are the volume and temperature of a gas sample related?

B. Mental Model—Draw a picture(s) that explains how the volume and temperature of a gas sample are related at the level of atoms and molecules and that illustrates the observations you made in the experiment. In words, explain how your picture(s) illustrate this relationship.

PRESSURE OF VAPORS—Exp. E-2

Name_________________________________ Lab Section ____________________________

Lab Partner____________________________

Problem Statements: How are vapor pressures and temperature related?
How are the vapor pressures of different liquids related?

I. Data Collection: The Apparatus

A. Obtain and calibrate a pressure-measuring device, a clean and dry 125 mL suction flask, and other materials needed to set up the apparatus shown in the figure below.

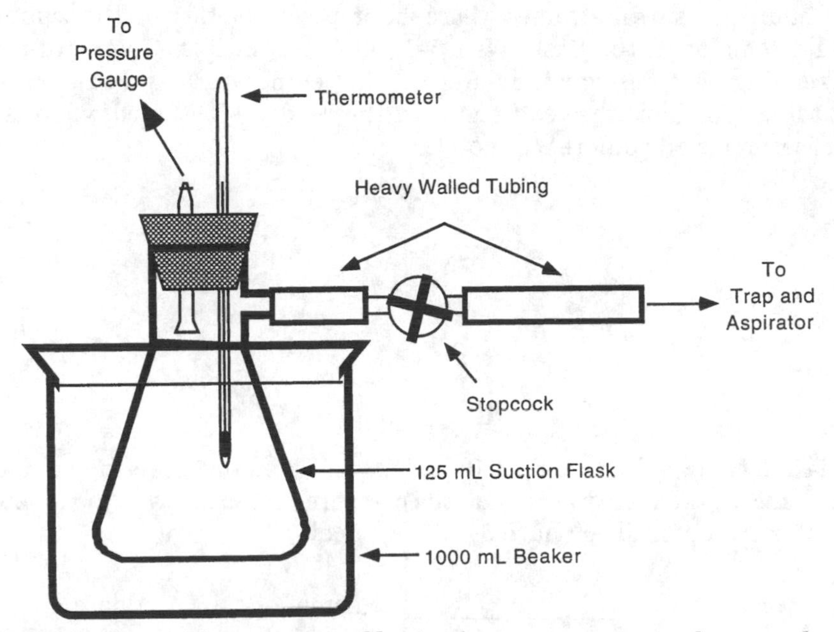

B. Fill the 1000 mL beaker with tap water. Obtain the temperature and atmospheric pressure and record these data in the table on the next page. Open the stopcock and turn on the aspirator until no further change takes place. Close the stopcock and then turn off the aspirator. Test the apparatus for leaks. If no change occurs for several minutes, proceed. If not, see your instructor.

C. Describe what happened. Compare the pressure reading with atmospheric pressure.

D. Calculate the pressure in the flask. Show your calculations below and record the result in the following table.

II. Data Collection: Hexane

CAUTION!!! Hexane is a flammable liquid. Keep it away from open flames.

A. Disconnect the aspirator and open the stopcock. Open the suction flask and add 15 mL of hexane. Reconnect the apparatus, close the stopcock, and turn on the aspirator. Open and close the stopcock until a minimum pressure is attained. Close the stopcock and turn off the aspirator. Make sure some of the liquid remains in the flask. Wait 5–10 minutes until the temperature stabilizes and the pressure reading becomes constant. Record the temperature and the pressure reading in the following table. Compare the readings with the readings obtained when the flask was empty. Propose a reason for any differences noticed.

B. Cool the flask by adding ice to the 1000 mL beaker. When the ice has melted and the temperature is stabilized, record the temperature and the pressure-reading data in the following table. Continue cooling and recording until you have 6 readings in all.

Atmospheric pressure __________ at __________ (time), __________ (date)

Pressure inside empty flask __________ at __________ (temp)

Hexane

Temperature	Pressure Reading
______________	______________
______________	______________
______________	______________
______________	______________
______________	______________
______________	______________

C. Disconnect the aspirator, open the stopcock, pour out the hexane, connect the system and evacuate it dry.

III. Data Analysis: Hexane

A. Determine the pressure exerted by the vapor of the hexane for each temperature and record these data in the following table. Show the calculations for the first data point below.

Hexane

Temperature	Vapor Pressure
______________	______________
______________	______________
______________	______________
______________	______________
______________	______________
______________	______________

B. What patterns are shown in these data? It might be helpful to graph the data.

IV. Data Collection: Methanol and Acetone

A. Repeat part II with methanol and acetone. (**CAUTION!!!** Methanol and acetone are flammable liquids. Keep them away from open flames.)

Methanol

Pressure Reading	Pressure	Temperature

Acetone

Pressure Reading	Pressure	Temperature

V. Data Analysis

Discuss and compare the vapor pressure plots for the three liquids.

VI. Interpretation

A. Mental Model—Propose a model that explains how a liquid develops a specific gaseous pressure. Draw a picture showing your explanation.

B. Explain how and why temperature affects the vapor pressure of a liquid.

C. Explain the relative values of the vapor pressures of the three substances you studied in this experiment. Predict how the vapor pressure of water would compare with these three substances. Justify your prediction. It might be helpful to draw the structures of these molecules and to consider their size, shape, and intermolecular attractions.

CHEMICAL PROPERTIES—Exp. F-1

Name______________________________ Lab Section __________________________

Lab Partner__________________________

Problem Statement: How do chemical properties of elements coordinate to their position in the periodic table?

I. General Instructions

A. **CAUTION!!!** Most of the chemicals you will be using can cause severe burns to flesh and clothing. This is particularly true of Na or K metal; do not let these come in contact with your person. Handle all solid chemicals with forceps, tongs or a spatula, ***not with your fingers.***

Take care that none of the solutions you create or take from the reagent shelf come into contact with your person or are spilled on the lab bench. If this does occur, rinse the affected area generously with water.

B. Chemical Facts Needed

1. See Table C-2 in Appendix C for properties of gases such as H_2, O_2, and Cl_2.
2. An acid is any compound that will release H^+ ions in aqueous solution. A base is any compound that will release OH^- ions in aqueous solution. An amphoteric compound is one that will act either as an acid or as a base depending on what it reacts with.
3. pH papers can be used to see if a compound dissolved in water is acidic or basic. A basic compound that is insoluble in water will often dissolve in an acidic (i.e., HCl) solution. An acidic compound that is insoluble in water will often dissolve in a basic (i.e., NaOH) solution.

II. Data Collection and Analysis: Reactions with Water

A. Put about 150 mL of water into a 250 mL beaker. Fill a Pyrex test tube with water and invert it into the beaker so that little or no air is trapped in the top of the test tube. Using forceps or tongs, loosely wrap a piece of sodium metal in aluminum foil. Punch several holes in the foil to expose the lump of Na. Using forceps or tongs, immerse the wrapped sodium into the beaker of water and collect the evolved gas in the test tube. When the test tube is completely filled with gas, pull it out of solution, being careful to keep it in a vertical position with the open end down. Touch a lighted match to the open end and record your observations. Test the solution you have made with litmus paper or phenolphthalein and record your observations.

Write a balanced chemical equation for the reaction you have observed.

B. Repeat the experiment above using a chip of calcium metal (it need not be wrapped), and record your observations.

Write a balanced chemical equation for the reaction you have observed.

C. Explain how your observations support the chemical equations you have written.

D. In this experiment you are to observe the relative vigor with which various elements react with water. Fill three beakers about halfway with water, cover the beakers with watch glasses. Into one beaker place a piece of Na metal, K metal into another, and Ca metal into the third. Record your observations.

Repeat this procedure with a piece of Mg metal, Al metal, and a pinch of sulfur. Record your observations.

E. Is there any correlation between the reactivity of these elements and their position in the periodic table? Justify your answer from the data collected.

III. Data Collection and Analysis: Reactions with Acid

A. Put 75 mL of water into a 250 mL beaker, fill a Pyrex test tube with water, and invert it into the beaker as you did in part II.A. Add about 25 mL of 4.0M HCl solution to the beaker and mix. Using tongs, hold 2–3 pieces of Mg in the solution, collecting the gas in the test tube. Test the gas with a match as you did in part II.A. Record your observations.

Write a balanced chemical equation for the reaction you have observed.

Explain the reasoning you used in coming up with this equation.

B. Place 5–10 mL of 4.0M HCl solution into each of three test tubes. Drop a piece of Mg metal into one test tube, Al metal into another, and a pinch of sulphur into the third. Record your observations.

C. Is there a correlation between reactivity with acid and the position of the element in the periodic table? Justify your answer from the data collected.

IV. Data Collection and Analysis: Acid-Base Properties

In these experiments you will be dealing with compounds having the general formula EO_xH_y where x and y will be various whole numbers and E represents various elements. You are to determine whether these compounds are acid, base, neutral, or amphoteric compounds. (See part I.B.)

A. Dissolve a pellet of NaOH in about 50 mL of water. Test the solution to determine its acid-base properties. Record your results.

B. The compound MgO_2H_2 has been added to water for you. The compound is relatively insoluble in water. Shake the solution bottle and obtain about 5 mL of the slurry. Test the slurry to determine the acid-base properties of MgO_2H_2 by seeing how it reacts with HCl(aq); with NaOH(aq). Record your results.

C. The compounds COH_4, PO_4H_3, SO_4H_2, and ClO_4H also have been dissolved in water for you. Obtain about 5 mL of each solution. Test each for acid-base properties and record your observations and conclusions.

COH_4

PO_4H_3

SO_4H_2

ClO_4H

D. (Optional) Put 7 mL of 1 M $Al(NO_3)_3$ into a Pyrex test tube and add to it 3 mL of 6 M NH_4OH. Gently heat the test tube with a Bunsen burner for about 2 minutes (do not point the test tube at anyone). Allow the tube to cool, centrifuge it for about 3 minutes, then pour off the liquid. Fill the test tube halfway with distilled water, mix the solid with the water, centrifuge, and again pour off the liquid. The solid you have formed is AlO_3H_3.

With a stirring rod transfer a small amount of the solid to two test tubes. Add about 2 mL of 4 M HCl a drop at a time to one test tube and about 2 mL of 4 M NaOH a drop at a time to the other. Record your observations and your conclusions concerning the acid-base properties of AlO_3H_3.

E. Write the formulas of the ions that are present when the following compounds are dissolved in water.

$NaOH$

SO_4H_2

ClO_4H

COH_4

F. What conclusion can be drawn about the acid-base properties of EO_xH_y compounds and the position of the element, E, in the periodic table? Justify your answer using the data collected.

V. Interpretation

A. You have identified several individual trends in reactivity of the elements versus their positions in the periodic table (see parts II.E, III.C, and IV.F). Select one of these trends and develop a verbal explanation for it. Try to get down to the subatomic level here: What is happening to electrons (or electron clouds) in these elements to cause the trend?

B. Mental Model—Draw a picture illustrating your explanation above. That is, symbolize your ideas of what is happening to electrons by drawing comparative pictures of electrons, protons, and nuclei in two different species.

PROPERTIES OF HALOGENS—Exp. F-2

Name________________________________ Lab Section ______________________________

Lab Partner______________________________

Problem Statements: How do Halogens (Cl_2, Br_2, I_2) react with Halides (Cl^-, Br^-, I^-)?
Can we explain this reactivity?

I. Data Collection: Solubility of the Halogens and the Halides

A. Test the solubility of Cl^-, Br^-, and I^- in both water and cyclohexane (C_6H_{12}) by adding a small amount (enough to cover a match head) of NaCl, NaBr, and NaI to separate test tubes and adding 0.5–1.0 mL of water. Repeat the test with 0.5–1.0 mL of C_6H_{12}. Mix vigorously. Record your observations. What solubility patterns do you see?

B. Mix the Cl^- (in water) with the Cl^- (in C_6H_{12}). Mix the solutions vigorously for 15 seconds. Do the same for the Br^- solutions and I^- solutions. Record your observations. How do these results connect to those seen above?

C. Test the solubility of a few I_2 crystals in water and in C_6H_{12} in a manner similar to that in part I.A. Add about 1.0 mL of C_6H_{12} to the water solution of I_2 and mix vigorously. Record your observations. Contrast the solubility properties of I^- and I_2.

D. Obtain 10 mL each of water solutions of I_2, Br_2, and Cl_2. Place 5 drops of each in separate test tubes and add about 1 mL of C_6H_{12}. Mix vigorously. Repeat the test with 10 drops of each in about 1.0 mL of C_6H_{12}. Record your observations.

E. Summarize your observations on the colors and solubility of the halides and halogens in the following chart. This information will be used in section II to identify reaction products.

	Cl^-	Br^-	I^-	Cl_2	Br_2	I_2
in water						
in cyclohexane						

II. Data Collection: Reactions between the Halogens and Halides

A. Add 5 drops of Br_2 (in water) to 1 mL of Cl^- (in water). Mix thoroughly. Add 0.5 mL of C_6H_{12} and mix vigorously. Record your observations.

Does the cyclohexane test indicate a reaction between Br_2 and Cl^-? If so, write a balanced equation describing the reaction. If there was no reaction, write "no reaction." Defend your conclusion using the experimental data you collected.

B. Repeat the experiment substituting I^- (in water) for the Cl^-. Record your observations. Write and defend your equation.

C. Repeat parts II.A and B using Cl_2 (in water) with Br^- (in water) and with I^- (in water). Record your observations. Write and defend your equations.

D. Repeat parts II.A and B using I_2 (in water) with Cl^- (in water) and with Br^- (in water). Record your observations, Write and defend your equations.

III. Data Analysis

A. Summarize the data collected in section II. You might do this by comparing the reactivity of the three halogens and by comparing the reactivity of the three halides. Alternatively, you might set up a matrix of halogens against halides and indicate what pairs did or did not react.

B. What pattern exists in these data? Is there any correlation between the reactivity of the halogens and halides and their positions in the periodic table? Show how your answer fits the data collected. (To get started, you might predict if a reaction would occur between F_2 and Br^-; between At_2 and Br^-).

IV. Interpretation

A. Develop a verbal explanation for the pattern you identified in part III.B. Try to get down to the subatomic level here: What is happening to electrons (or electron clouds) in these atoms to cause the reactivity pattern? (To simplify the question, you can consider the halogens as separate neutral atoms; e.g., Br_2 as 2 Br.)

B. Mental Model—Illustrate your explanation. That is, symbolize your ideas of what is happening to electrons by drawing comparative pictures of electrons, protons, and nuclei in different species.

a. Mental Model—Draw a series of molecular pictures illustrating different phases of an acid base titration before, at, and after the end-point. Explain how your pictures account for your observations during the titration.

By definition:

(1) $\text{conc. of OH}^- \text{ solution} = \dfrac{\#\text{ moles of OH}^- \text{ added}}{\#\text{ liters of OH}^- \text{ solution}}$

The numerator of the right-hand term can be determined using the following reasoning:

Each molecule of the acid KHP can release one H^+ ion when dissolved in water according to the following equation:

$$C_6H_4(\text{C(=O)}-O^-)(\text{C(=O)}-O-H)\;\; K^+ \xrightarrow[H_2O]{\text{in}} H^+ + K^+ + C_6H_4(\text{C(=O)}-O^-)(\text{C(=O)}-O^-)$$

Therefore:

(2) # moles H^+ available in the compound = # moles KHP. The net reaction that is occurring as you titrate is: $H^+ + OH^- \rightarrow H_2O$. The OH^- added is consuming the H^+ being released by the compound. What is important is that at the end point of the titration:

(3) # moles OH^- added = # moles H^+ available in the compound.

MOLECULAR STRUCTURES—Exp. H-1

Pre-Lab Assignment: Study the sections of your textbook on Lewis Structures and Molecular Geometries (i.e., Valence-Shell Electron-Pair Repulsion theory). Alternatively or in addition, you can study the supplements to this experiment on pages 124–137 of this lab manual.

Study Supplement #3, page 138, on 3-D drawings.

In this laboratory period you will be asked to determine the Lewis Structures and the three-dimensional geometries of molecular species starting only with molecular formulas. Thus there will be no lab work, per se. The lab time is to be used to help you perfect your skills at deducing Lewis Structures and, most importantly, to help you become acquainted with various molecular geometries. Model kits or balls and sticks will be available so that you can construct 3-D models of the molecules.

You will be assigned to work with one of the lists of molecular species on the next page. You are to do the following for each species, writing your answers on the work sheets provided:

1. Determine the number of valence electrons in the molecule. (See your text and/or Supplement #1, page 124.)

2. Determine the Lewis Structure of the molecule. (See your text and/or Supplement #1, page 124.)

3. Determine the geometry around each centralized atom in the molecule from the Lewis Structure. See your text and/or Supplement #2, page 130.)

4. Construct an exact geometric model of the molecule using the materials provided.

5. Draw an exact 3-D representation of the molecule from the model. (See Supplement #3, page 138.)

Examples of how the work sheets are to be filled in are given on page 122.

MOLECULAR SPECIES LIST

List #1

1. $SiCl_4$ (no Cl-Cl bonds)
2. PCl_3 (no Cl-Cl bonds)
3. NO_3^- (no O-O bonds)
4. SF_6 (no F-F bonds)
5. PCl_5 (no Cl-Cl bonds)
6. IF_3 (no F-F bonds)
7. XeF_5^+ (no F-F bonds)
8. SO_4^{2-} (no O-O bonds)
9. C_4H_8O
10. C_5H_{12}

List #2

1. CH_3Br (C is central atom)
2. ICl_2^+ (no Cl-Cl bonds)
3. NO_2^- (no O-O bonds)
4. BF_3 (no F-F bonds)
5. HOOH (bonding as in formula)
6. $SbCl_5$ (no Cl-Cl bonds)
7. BrF_3 (no F-F bonds)
8. ICl_4^- (no Cl-Cl bonds)
9. $CH_3CH_2CO_2^-$ (bonding as in formula)
10. C_5H_{10}

List #3

1. PO_4^{3-} (no O-O bonds)
2. H_2NNH_2 (bonding as in formula)
3. CS_2 (no S-S bonds)
4. BCl_3 (no Cl-Cl bonds)
5. SbF_5 (no F-F bonds)
6. PCl_6^- (no Cl-Cl bonds)
7. ClF_5 (no F-F bonds)
8. XeF_2 (no F-F bonds)
9. C_3H_5Cl (C's are central atoms)
10. C_6H_{12} (6 C's in a ring)

List #4

1. NH_4^+
2. SF_2 (no F-F bonds)
3. COF_2 (C is central atom)
4. SO_2 (no O-O bonds)
5. PBr_5 (no Br-Br bonds)
6. $TeCl_4$ (no Cl-Cl bonds)
7. XeF_4 (no F-F bonds)
8. ClO_4^- (no O-O bonds)
9. C_3H_8O
10. C_6H_6 (6 C's in a ring)

MOLECULAR STRUCTURES

Name_______________________ Section_______________________

Partner _______________________ Molecule List # _______________________

1. # valence electrons = _______________

Lewis Structure: Molecular geometry = _______________

3-D Drawing:

2. # valence electrons = _______________

Lewis Structure: Molecular geometry = _______________

3-D Drawing:

3.

Lewis Structure:

valence electrons = ______________________

Molecular geometry = ______________________

3-D Drawing:

4.

Lewis Structure:

valence electrons = ______________________

Molecular geometry = ______________________

3-D Drawing:

5.

Lewis Structure:

valence electrons = ____________________

Molecular geometry = ____________________

3-D Drawing:

6.

Lewis Structure:

valence electrons = ____________________

Molecular geometry = ____________________

3-D Drawing:

7.

valence electrons = ______________________

Lewis Structure:

Molecular geometry = ______________________

3-D Drawing:

8.

valence electrons = ______________________

Lewis Structure:

Molecular geometry = ______________________

3-D Drawing:

9.

Lewis Structure:

valence electrons = ____________

Molecular geometry = ____________

3-D Drawing:

10.

Lewis Structure:

valence electrons = ____________

Molecular geometry = ____________

3-D Drawing:

Examples of Work to be Done During the Lab Period

(Note: Only the molecular formulas will be given, you must determine the number of valence electrons, the Lewis Structure, the geometry and give a 3-D drawing.)

1. NH_3

\# valence electrons = 8

Lewis Structure:

Molecular Geometry = triangular pyramid

```
H—N̈—H
   |
   H
```

3-D Drawing:

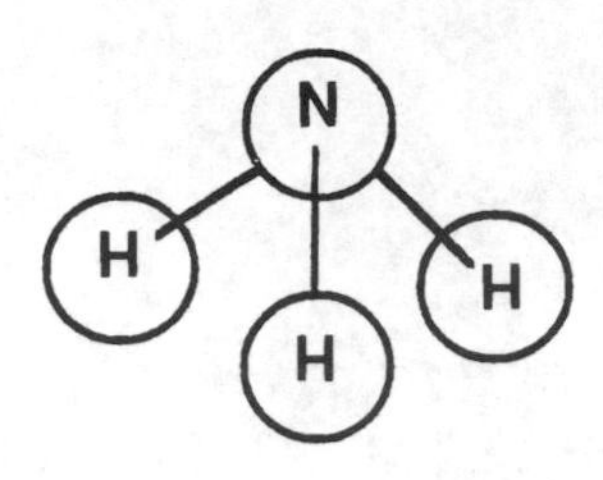

2. $CH_3CH_2CH_3$

\# valence electrons = 20

Lewis Structure:

molecular Geometry = tetrahedral about each C atom

```
   H  H  H
   |  |  |
H—C—C—C—H
   |  |  |
   H  H  H
```

3-D Drawing:

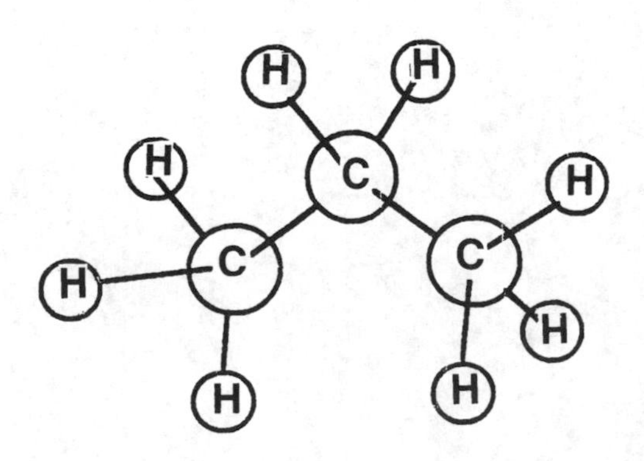

SUPPLEMENTS ON MOLECULAR STRUCTURES

The following pages contain study supplements on the various topics involved in determining molecular structures. Your textbook and the classroom lectures may also provide information on these topics.

The supplements contain activities designed so that you can interact with them. The activities are divided into sections designated by a capital letter. Most of these sections ask questions which you should answer by writing in the space provided. The sections are designed so that you can compare your answer with the correct answer before proceeding to the next section. These answers are designated by lower case letters.

Make a mask by folding a sheet of paper. Cover the sections below the one you are studying. When you have answered the question, slide the mask down to expose the correct answer and compare it with yours. If it is the same or you understand your error, proceed to the next section. If you don't understand your error, see your instructor.

The supplements are designed so that you can study (and refer back to) just those topics that are giving you difficulty.

Supl. #1—LEWIS STRUCTURES

a. Definitions and Ideas

The distinguishing characteristic of a molecule or molecular ion is that its atoms are held together by what are called *covalent bonds*. In covalent bonding two atoms are attracted to the same set of electrons. This mutual attraction to the same electrons keeps the atoms a set distance apart. The mutual attraction, in effect, bonds the atoms together. The two atoms are said to *share* the set of *bonding electrons*. The three principle types of covalent bonds are symbolized below using carbon atoms.

C—C	C = C	C ≡ C
single bond	double bond	triple bond
2 electrons shared	4 electrons shared	6 electrons shared

The only electrons that can become involved in covalent bonding are those in the outer energy shell (highest n level) of the atom. (The other electrons, those in the inner energy levels, are simply buried too deeply in the atom to be affected by an outside nucleus.) The outer shell electrons of an atom are called its *valence electrons*. In a molecule or molecular ion, some of these valence electrons will be shared in covalent bonds. Other valence electrons, called *unshared pairs*, will be attracted to only one atom.

A *Lewis Structure* is a line-and-dot picture that shows where the valence electrons are and what they are doing in a molecule. The Lewis Structure for acetic acid, CH_3CO_2H, is shown below.

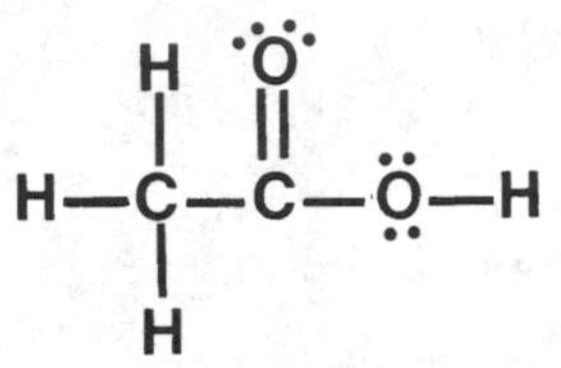

```
    H  :O:
    |  ||
 H—C—C—Ö—H
    |     ..
    H
```

It shows, for example, that each hydrogen atom is held in place by a single bond and has no unshared valence electrons on it; that one oxygen has two pairs of unshared valence electrons and is sharing four others with a carbon atom (is double bonded).

The first Lewis Structures had to be determined by interpreting experimental data on the molecules. However, enough of this work has been done to enable us to *predict* Lewis Structures with reasonable certainty. In other words, the principles used by nature to distribute valence electrons in molecules have been discovered.

Controlling Ideas for Predicting Lewis Structures

There are many methods for deducing Lewis Structures. The method used in the remaining parts of this supplement may differ from the one given in your textbook. Whatever method is used, the following ideas are controlling the work.

1. **Number of Valence Electrons**. The Lewis Structure picture must show just the number of valence electrons used in the actual molecule or molecular ion.
2. **Octet (duet) Rule**. Each atom in the molecule will usually have 8 valence electrons on it (octet rule). This counts *all* the valence electrons the atom is sharing as well as its unshared pairs. Hydrogen atoms will have only 2 valence electrons (duet rule) each. (You will see that the octet rule is not always followed by the atoms in periods 3 and higher of the periodic table.)

Note: The method for deducing Lewis Structures given in the remainder of this supplement only works well with covalent compounds containing "A" Group elements; i.e., representative elements.

b. The Number of Valence Electrons

	A. Generate the electron configuration for each different atom in a $HCCl_3$ molecule. H = C = Cl =
a. H = $1s^1$ C = $1s^2 2s^2 2p^2$ Cl = $1s^2 2s^2 2p^6 3s^2 3p^5$	B. The number of valence electrons an atom has can be determined in several ways. One is to generate the electron configuration of the atom and count the number of electrons in the highest n level. (These will be the outermost s and p orbital electrons.) Determine the number of valence electrons for each atom in the molecule $HCCl_3$. H = C = Cl =
b. H = 1 C = 4 Cl = 7	C. The total number of valence electrons in a neutral molecule can be determined by summing the number of valence electrons for each atom in the molecule. If the species is a molecular ion, one must go one step further to account for the charge. Add the extra electrons if it is an anion (negative ion) or subtract the lost electrons if it is a cation (positive ion). Determine the total number of valence electrons in a $HCCl_3$ molecule. (Remember there are three Cl atoms.) # valence electrons = ________
c. # val. e^- = 26 1 H x 1 = 1 1 C x 4 = 4 3 Cl x 7 = 21 26	D. Thus 26 electrons must be shown in the Lewis Structure of $HCCl_3$. For other examples of calculating the # valence electrons see the three Lewis Structure examples on the next pages.

c. Determining Lewis Structures

Example #1

	A. You are to determine the Lewis Structure for PCl_3, given only its molecular formula and the fact that there are no Cl-Cl bonds. What does this last fact tell you?
a. Each Cl is individually bonded to the P atom.	B. Determine the total number of valence electrons in a PCl_3 molecule. # valence electrons = ____________
b. # val. e^- = 26 1 P x 5 = 5 3 Cl x 7 = 21 26	C. We will ignore this number for the moment and just guess at a possible Lewis Structure. Recalling answer a, connect the appropriate atoms with single bonds.
c. Cl—P—Cl | Cl	D. Now add unshared electron *pairs* to each atom until each has 8 (shared plus unshared) electrons on it. If hydrogen atoms were present, each would only have 2 electrons.
d. :C̤̈l—P̈—C̤̈l: | :C̤̈l:	E. This is our "first guess" at the Lewis Structure and must be checked for number of electrons. How many electrons are shown in the picture?
e. 26 e^- 10 unshared pairs 3 bonded pairs = 13 pairs = 26 e^-	F. This number is exactly what we calculated that nature uses in the molecule (# valence electrons). Since our "first guess" Lewis Structure contains the correct number of electrons and since each atom has an octet, this is a correct Lewis Structure for PCl_3.

Example #2

A. You are to determine the Lewis Structure for the molecular ion CO_3^{2-} (no O-O bonds).

Determine the total number of valence electrons in one CO_3^{2-} ion.

valence electrons = ____________________

---

a. # val. e^- = 24

1 C x 4 = 4

3 O x 6 = 18

extra e^- = 2

24

B. The species is charged because it has 2 more electrons than those contributed by the neutral atoms.

Recalling that there are no O-O bonds, draw a picture connecting the appropriate atoms with single bonds.

---

b.
```
O—C—O
  |
  O
```

C. Now add unshared electron pairs until each non-hydrogen atom has a total of 8.

---

c.
```
 ..  ..  ..
:O — C — O:
 ..  |   ..
    :O:
     ..
```

D. Count the number of electrons in this "first guess" picture.

How does this number compare to the calculated # val. e^- (ans. a)?

---

d. 26 e^- shown. This is 2 e^- too many, since 24 e^- are required.

E. Two electrons must be removed from our "first guess" picture. Remove them as an unshared pair from the central atom; i.e., the carbon atom.

e. :Ö—C—Ö:
      |
    :Ö:
	F. This structure has the required 24 electrons but now the C atom has only 6 electrons instead of the required 8.
	
	This problem is solved by turning an unshared pair on one of the oxygens into a bonded pair. That is, by using an unshared pair to create a double bond,
	
	i.e., —C—Ö:
	
	Redraw the picture using this idea.

f. :Ö—C=O:
      |
    :Ö:
	G. This Lewis Structure contains the required 24 electrons and each atom has an octet. Thus it is a correct Lewis Structure.

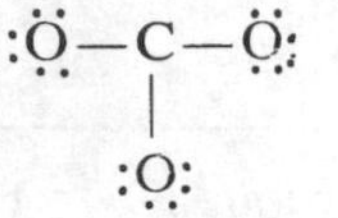

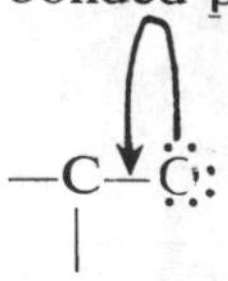

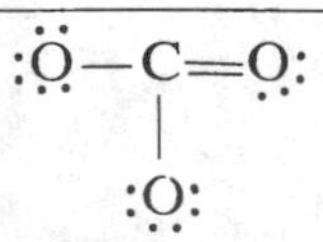

In general, whenever your "first guess" picture contains too many electrons, the correct Lewis Structure will contain double or triple bonds. (These so called multiple bonds occur most often between atoms of C, N, O, P, S, Se. These six multiple bonders are easy to recall because they form a triangle in the periodic table.)

A molecule where the multiple bond can be located in two or more positions is best represented by a series of Lewis Structures that is given the term *resonance*. Consult your text or your instructor for details.

Example # 3

	A. You are to determine the Lewis Structure for the molecule SF_4 (no F-F bonds). Determine the total number of valence electrons in one SF_4 molecule. # valence electrons = ________________
a. # val. e^- = 34 1 S x 6 = 6 4 F x 7 = 28 34	B. Draw a picture connecting the appropriate atoms with single bonds. Recall that SF_4 has no F-F bonds.

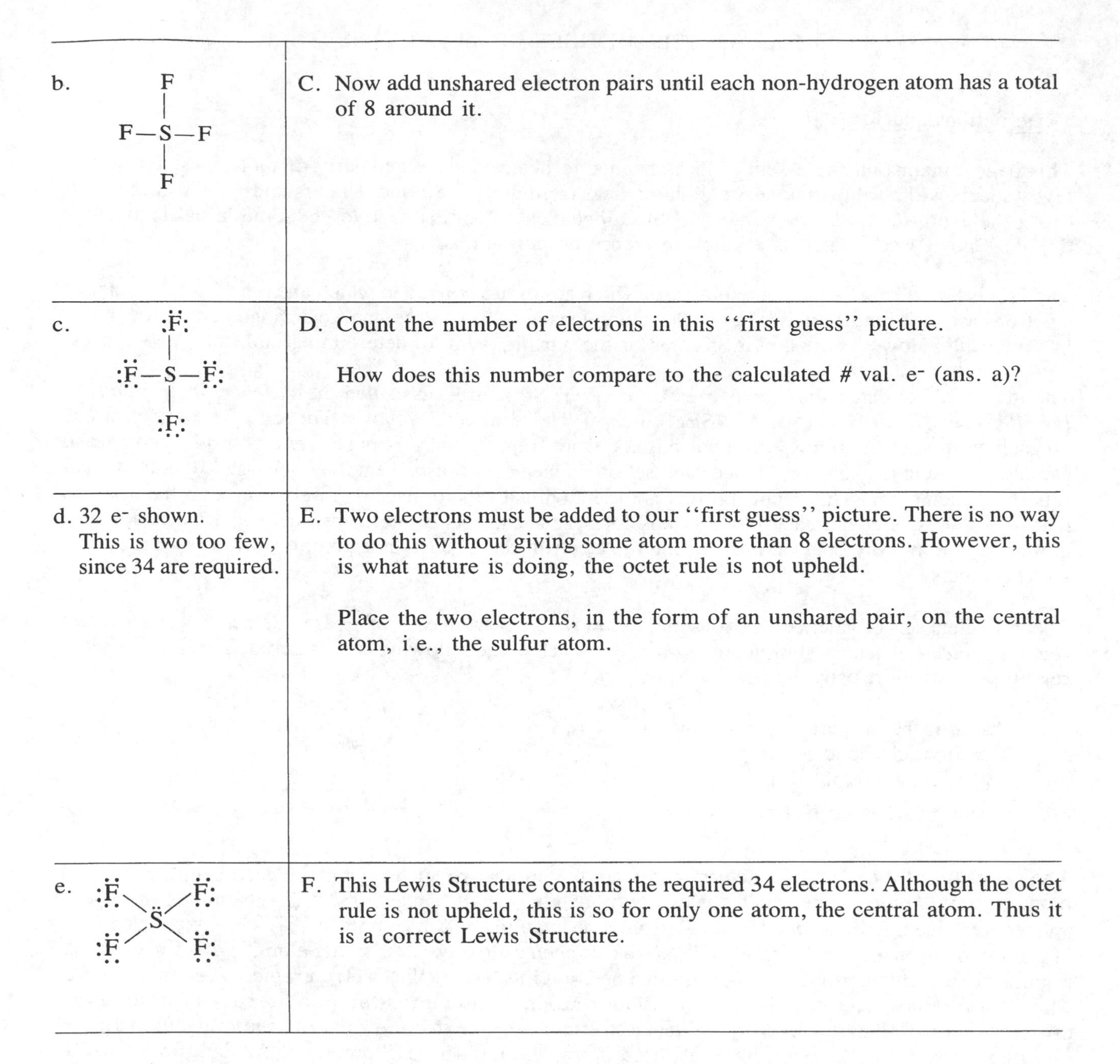

b. F–S–F with F above and F below S (four F atoms single-bonded to S)	C. Now add unshared electron pairs until each non-hydrogen atom has a total of 8 around it.
c. :F̤: above, :F̤–S–F̤: , :F̤: below (each F with three unshared pairs)	D. Count the number of electrons in this "first guess" picture. How does this number compare to the calculated # val. e^- (ans. a)?
d. 32 e^- shown. This is two too few, since 34 are required.	E. Two electrons must be added to our "first guess" picture. There is no way to do this without giving some atom more than 8 electrons. However, this is what nature is doing, the octet rule is not upheld. Place the two electrons, in the form of an unshared pair, on the central atom, i.e., the sulfur atom.
e. Four :F̤: atoms bonded to S̈ (unshared pair on S)	F. This Lewis Structure contains the required 34 electrons. Although the octet rule is not upheld, this is so for only one atom, the central atom. Thus it is a correct Lewis Structure.

In general whenever your "first guess" picture contains too few electrons, the octet rule will be broken on the central atom. (The central atom will be in the third or higher period of the Table. Such atoms have d orbitals of low enough energy to take the "extra" electron pairs.)

Note that multiple bonds cannot help in such "too few electrons" cases.

Supl. #2—GEOMETRIES OF MOLECULES

a. Definitions and Ideas

This page contains all the essential ideas needed to determine the geometry of molecules. However, few students will be able to understand these ideas from this page alone. Understanding will come from doing the exercises in the remainder of this supplement. Use this page to get some initial feeling for the ideas now. Use it later as a quick reference on definitions.

The Lewis Structure of a molecule indicates which atoms are bonded to which and where all the valence electrons are in the molecule. A Lewis Structure does not show the geometric arrangement of the atoms in a molecule. However the Lewis Structure is the starting point for determining molecular geometries.

The ideas used to determine geometries from Lewis Structures are called the *Valence-Shell Electron-Pair Repulsion Theory* (abbreviated VSEPR theory). The ideas are as follows. The sets of valence electrons on each atom are free to move around, but not away from, the atom (much like a dog on a short leash can move around his handler). Since each set of valence electrons is negatively charged, the individual sets repel each other. This repulsion forces the individual sets to place themselves around the atom so that they are as far from each other as possible. Thus the sets of electrons occupy a fixed geometry around the atom. Any other atoms sharing the electron sets (i.e., bonded atoms) will also occupy this fixed geometry.

The individual sets of valence electrons that can move away from each other are often called *independent regions of electron density*. For geometry purposes each of the following is considered as one independent region of electron density:

- .., each unshared pair is 1 region
- —, each single bond is 1 region
- =, each double bond is 1 region
- ≡, each triple bond is 1 region

The geometry of a molecule is determined from the number of such regions on each centrally located atom. A *central atom* is one which has two or more other atoms bonded to it. A given number of regions on a central atom will assume the fixed geometry given in the table on page 137. This fixed geometry of the regions is sometimes called the *electron geometry*. Bonded atoms "tag along" with the electron regions of the central atom, taking the same position in space as the electron region they are sharing. The *molecular geometry* describes the positions the bonded atoms take around the central atom. If the central atom has all of its electron regions used in bonding, the electron and molecular geometries are identical. If the central atom has unshared pairs, the molecular geometry is a "daughter" of the electron geometry as shown in the last column in the table on page 137.

b. VSEPR Theory

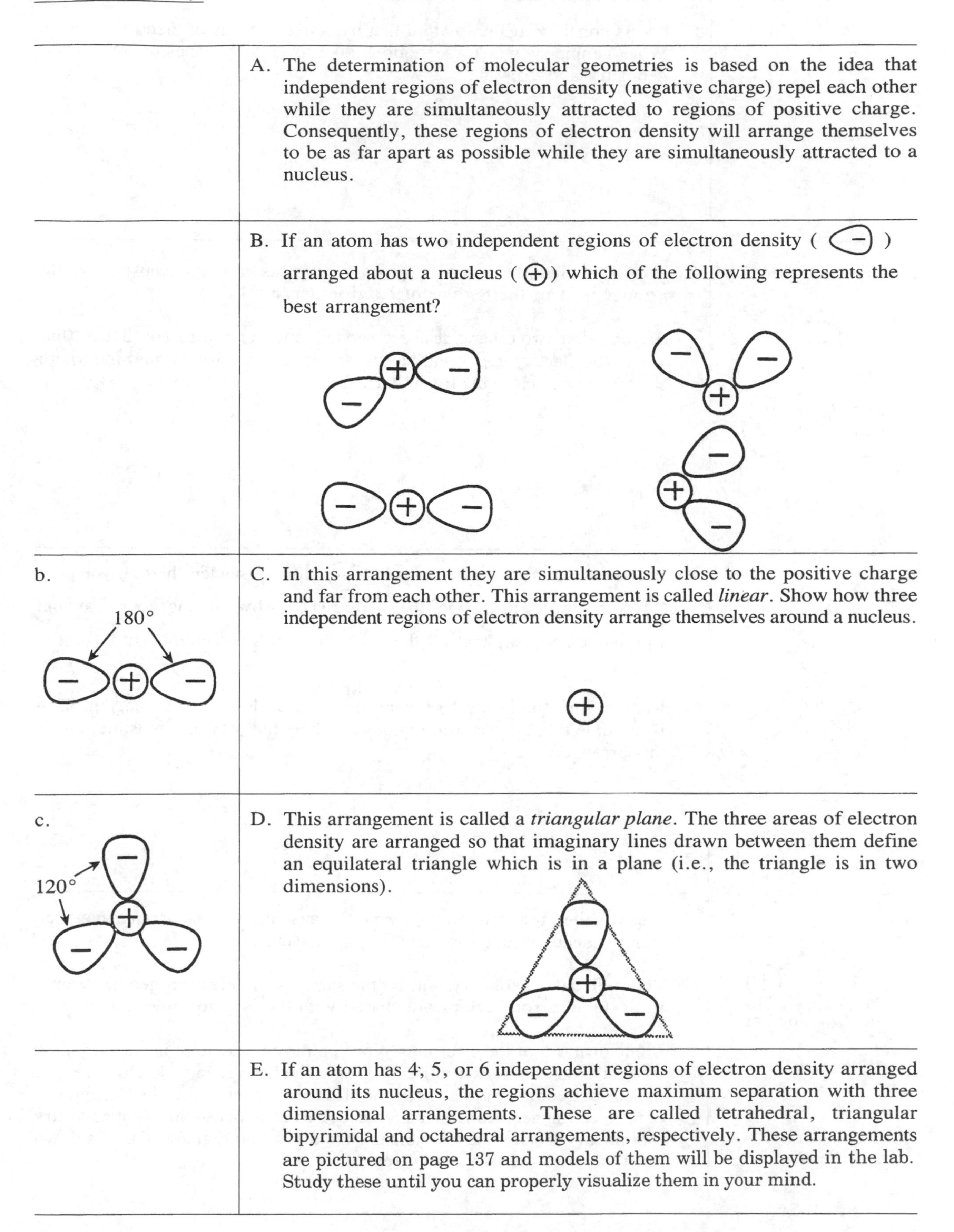

	A. The determination of molecular geometries is based on the idea that independent regions of electron density (negative charge) repel each other while they are simultaneously attracted to regions of positive charge. Consequently, these regions of electron density will arrange themselves to be as far apart as possible while they are simultaneously attracted to a nucleus.
	B. If an atom has two independent regions of electron density (−) arranged about a nucleus (+) which of the following represents the best arrangement?
b. 180°	C. In this arrangement they are simultaneously close to the positive charge and far from each other. This arrangement is called *linear*. Show how three independent regions of electron density arrange themselves around a nucleus.
c. 120°	D. This arrangement is called a *triangular plane*. The three areas of electron density are arranged so that imaginary lines drawn between them define an equilateral triangle which is in a plane (i.e., the triangle is in two dimensions).
	E. If an atom has 4, 5, or 6 independent regions of electron density arranged around its nucleus, the regions achieve maximum separation with three dimensional arrangements. These are called tetrahedral, triangular bipyrimidal and octahedral arrangements, respectively. These arrangements are pictured on page 137 and models of them will be displayed in the lab. Study these until you can properly visualize them in your mind.

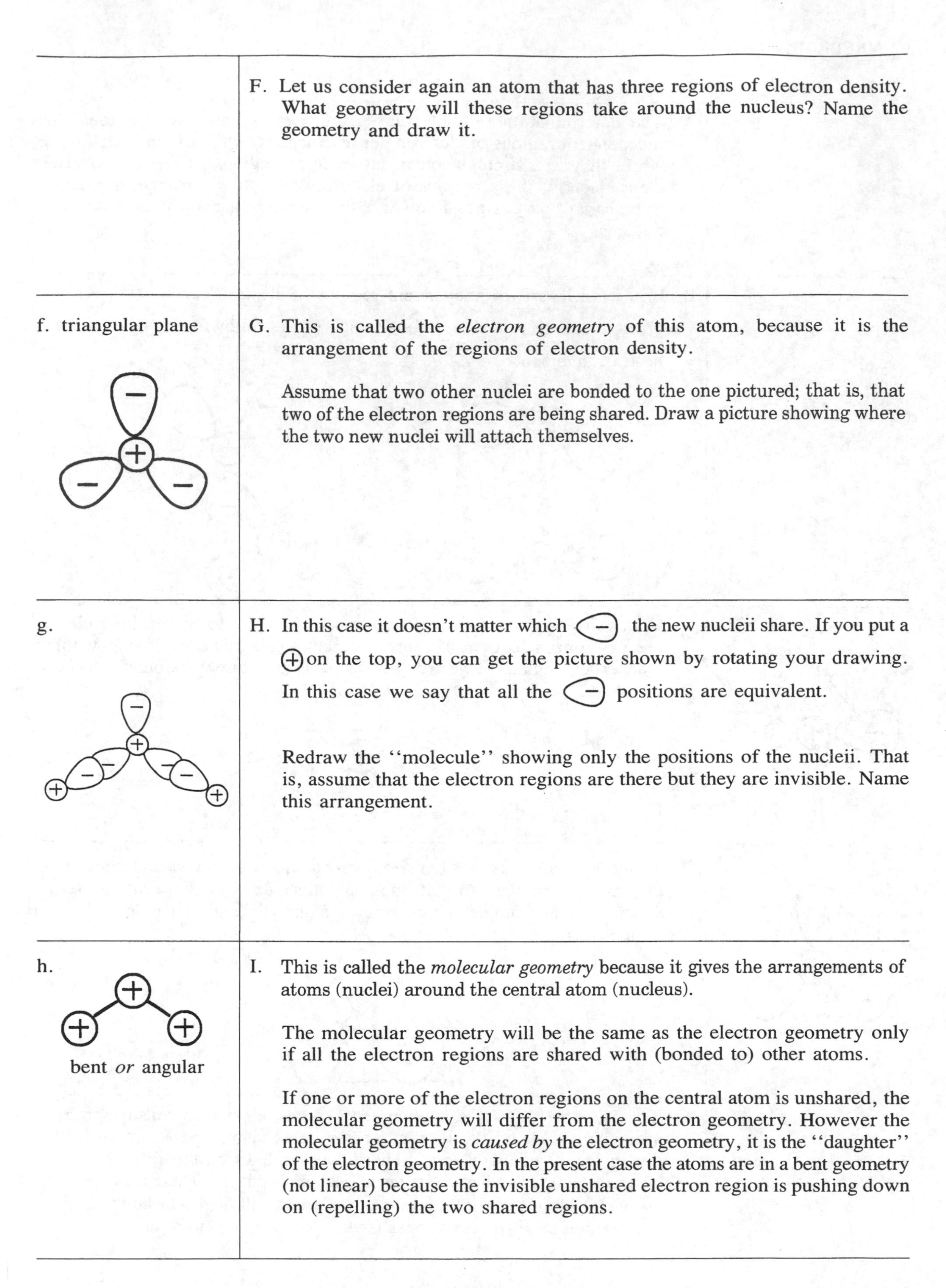

F. Let us consider again an atom that has three regions of electron density. What geometry will these regions take around the nucleus? Name the geometry and draw it.

f. triangular plane

G. This is called the *electron geometry* of this atom, because it is the arrangement of the regions of electron density.

Assume that two other nuclei are bonded to the one pictured; that is, that two of the electron regions are being shared. Draw a picture showing where the two new nuclei will attach themselves.

g.

H. In this case it doesn't matter which ⊖ the new nucleii share. If you put a ⊕ on the top, you can get the picture shown by rotating your drawing. In this case we say that all the ⊖ positions are equivalent.

Redraw the ''molecule'' showing only the positions of the nucleii. That is, assume that the electron regions are there but they are invisible. Name this arrangement.

h.

bent *or* angular

I. This is called the *molecular geometry* because it gives the arrangements of atoms (nuclei) around the central atom (nucleus).

The molecular geometry will be the same as the electron geometry only if all the electron regions are shared with (bonded to) other atoms.

If one or more of the electron regions on the central atom is unshared, the molecular geometry will differ from the electron geometry. However the molecular geometry is *caused by* the electron geometry, it is the ''daughter'' of the electron geometry. In the present case the atoms are in a bent geometry (not linear) because the invisible unshared electron region is pushing down on (repelling) the two shared regions.

c. Determining Molecular Geometries

Example # 1

	A. You are to determine the molecular geometry of CO_3^{2-} (no O-O bonds). Determine the Lewis Structure of the molecular ion.
a. :Ö—C=Ö with :Ö: bonded below C	B. Which is the central atom in the molecule? central atom = ________
b. The C atom	C. How many independent regions of electron density are on this central atom? # regions = ________
c. 3 regions	D. Note that the double bond counts as only one region because the 4 electrons in the double bond must stay together. How will the 3 regions arrange themselves around the carbon? That is, what it the electron geometry? (Consult the Table on page 137.) electron geometry = ________
d. triangular plane	E. How many of the three regions are used in bonding to other atoms? # bonded regions = ________
e. all 3	F. What will be the arrangement of oxygen atoms around the carbon? What is the molecular geometry? molecular geometry = ________
f. triangular plane (O, C, O, O diagram)	G. The molecular geometry is identical to the electron geometry because all the electron regions on the central atom are shared (bonding) electron regions.

Example # 2

	A. You are to determine the molecular geometry of SF_4 (no F-F bonds). Determine the Lewis Structure of the molecule.
a. 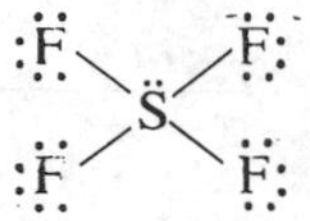 :F̤ F̤: / S̈ / :F̤ F̤:	B. Which is the central atom in the molecule? central atom = ________
b. the S atom	C. How many independent regions of electron density are on this central atom? # regions ________
c. 5 regions. 1 unshared pair + 4 single bonds	D. According to the table on page 137, how will these 5 regions arrange themselves around the S atom? electron geometry = ________
d. triangular bipyramid	E. To properly see this geometry you should inspect the model on display in the lab or build your own with the materials provided. The geometry has 3 regions in a triangular plane (t regions). The other two are directly above and below the triangle, forming the apex (a regions) of two pyramids. 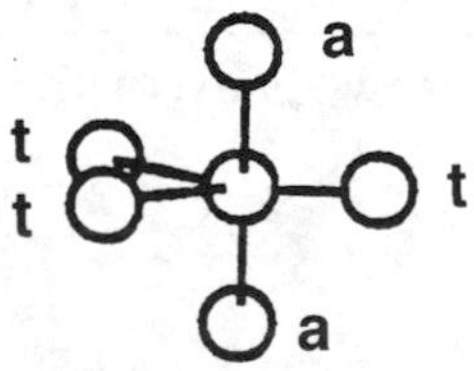 a t t t a How many of the 5 regions are used in bonding? # bonded regions = ________
e. 4	F. Since not all regions are used in bonding, the molecular geometry will not be the same as the electron geometry but will be some derivative of it. We must determine where the unshared pair will be located in the electron geometry. Refer to the figure in E above. If the unshared pair locates in one of the apex (a) positions, the molecular geometry would be a (single) triangular pyramid. If the unshared pair locates in one of the triangle (t) positions, a seesaw molecular geometry is obtained. (The other two t positions would be the legs; the a positions, the seats of the seesaw.)

	G. According to the last column in the table on page 137, which of the possible geometries is actually used in nature? molecular geometry = __________
g. seesaw	

Example #3

	A. You are to determine the molecular geometry of a methylamine molecule, CH_3NH_2. Determine the Lewis Structure of the molecule (the H atom bonding is indicated by the formula).
a. H–C(H)(H)–N̈(H)–H	B. Which is (are) the central atom(s) in the molecule? central atom(s) = __________
b. both the C and the N atoms	C. In cases where there is more than one central atom, the total geometry depends on the geometry around each central atom. We will apply the information in the table on page 137 to each central atom individually and put the two geometries together only in a final step. How many independent regions of electron density are on each central atom? # regions: C = __________ N = __________
c. C = 4 N = 4	D. Note that the bond between the C and the N is counted twice, as a region belonging to each atom. How will the regions around each central atom arrange themselves? electron geometries: C = __________ N = __________

d. C = tetrahedron N = tetrahedron	E. How many of the regions around each central atom are used in bonding? # bonded regions: C = __________ N = __________
e. C = 4 N = 3	F. According to the Table what will be the arrangement of atoms around each central atom? molecular geometries: C = __________ N = __________
f. C = tetrahedron N = triangular pyramid triangular pyramid 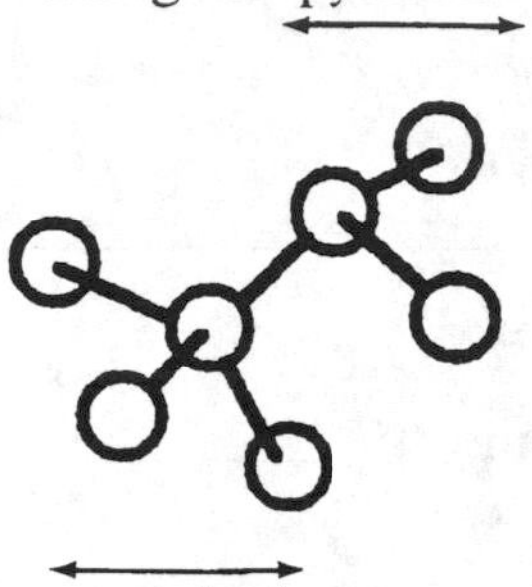 tetrahedron	G. We are now ready to put these individual geometries together to form the molecule. Naming the total molecular geometry is best done by just stating the individual geometries around each central atom; i.e., by giving answer f. Making a model of the molecule is a bit more involved. The molecule is a tetrahedron sharing an atom with a triangular pyramid. Make separately a tetrahedron and a triangular pyramid. Now remove a ''bond'' stick from one model and insert a ''bond'' stick from the other model in its place. The result is illustrated.

Table of Geometries

Number of regions of electron density	Arrangement of regions for maximum separation	Number of regions used in bonding	Geometry of molecule
2	linear 	2	linear
		1	linear
3	triangular plane 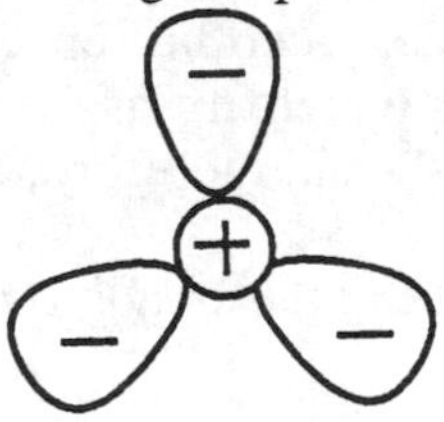	3	triangular plane
		2	angular
		1	linear
4	tetrahedron 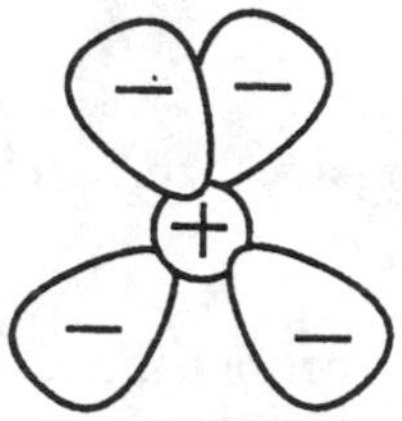	4	tetrahedron
		3	triangular pyramid
		2	angular
		1	linear
5	triangular bipyramid 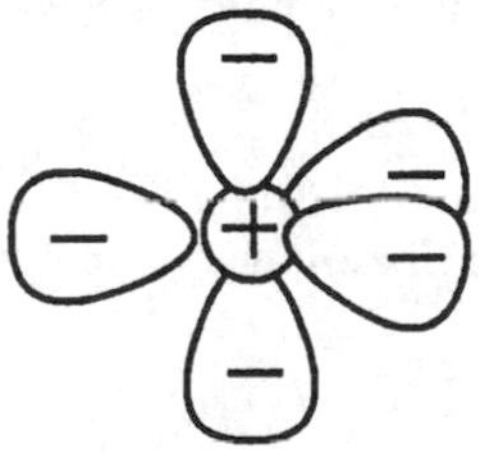	5	triangular bipyramid
		4	sea saw
		3	T
		2	linear
6	octahedron 	6	octahedron
		5	square pyramid
		4	square plane

Often it is necessary to try to represent three dimensional molecular geometries in drawings. The following instructions are intended to illustrate a method for drawing 3-D geometries. The important consideration is to obtain realistic drawings which represent the geometry of the molecule.

A. In order to keep the situation simple, atoms in a molecule will be represented by circles. Smaller atoms can be represented by smaller circles. The bond attaching atoms will be represented by lines connecting the atoms. For example, the molecule HCl can be represented as:

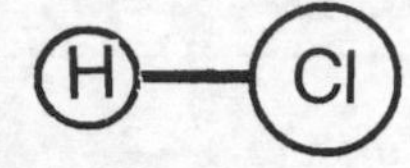

B. Molecules with geometries which are in one plane can be represented on paper quite easily. The difficulty arises with some of the molecules which are based on the tetrahedron, triangular bipyramid and octahedron. Some of these are 3-D and therefore, require perspective drawings.

C. A technique for suggesting perspective is to show the bonds overlapping the circle which is behind the plane of the paper.

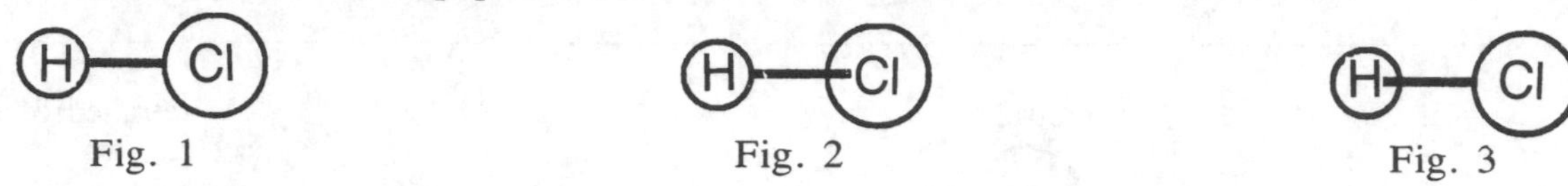

Fig. 1 Fig. 2 Fig. 3

In figure 1 both Cl and H are in the plane of the paper. In figure 2 the Cl is behind the H and in figure 3 Cl is in front of H.

D. Consider the molecule CH_4 which has a Lewis Structure of bonded

```
   H
   |
H—C—H.
   |
   H
```

Since there are 4 electron pairs around the C atom and no non-bonded pairs, the geometry around the C atom must be tetrahedral. A 3-D drawing of this geometry is obtained as follows. (See the table on page 137.)

E. Assume you have made a 3-D ball and stick model for CH_4. You rotate it in your hand until the central atom and one of the hydrogens are in one plane. A drawing of that might look like this:

F. If this orientation is rotated until two other hydrogens are in front of the plane and one is behind, the placement of circles might look like this:

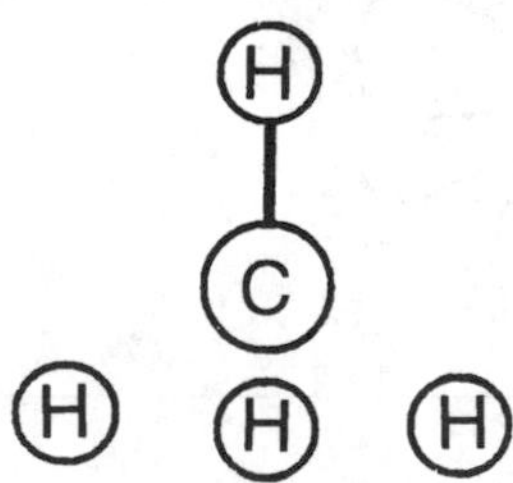

G. Then if lines were drawn so that the two in front overlap the carbon and the one in back overlaps one hydrogen, the final drawing looks like this:

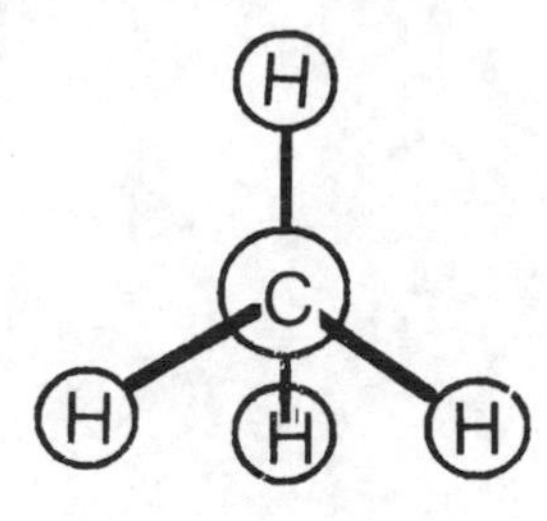

H. The following steps are suggested for deriving 3-D drawings:

(1) Build a 3-D model of the molecule.
(2) Rotate the model until as many atoms as possible are in the same plane as the central atom. Make sure all the atoms are showing.
(3) Draw circles representing the placement of all atoms in the molecule.
(4) By considering the central atom to be in the plane, draw all the atoms as being in the same plane, in front of, or behind the central atom by overlapping bond lines.

MOLECULAR MODELS ON A COMPUTER—Exp. H-2

Name__________________________________ Lab Section ______________________________

Lab Partner______________________________

In this laboratory activity you will examine three-dimensional models of molecules using a computer-based molecular viewing program. The laboratory activities are divided into exercises covering several topics. In each exercise you will be asked to compare combinations of molecules and to answer questions about them. Your laboratory instructor will specify which exercises you will do. In order to do this laboratory you will need text files that describe the molecules you will examine and a computer program that can translate those text files into computer graphics of the molecules. Your laboratory instructor will provide, or inform you how to obtain, the molecule text files, the computer viewing program, and instructions on how to manipulate the molecules on your computer screen.

I. VSEPR

(Refer to the table of geometries in experiment H-1, page 137.)

A. Examine the molecules with two regions of electron density. Draw the molecules and measure their bond angles.

B. Examine the molecules with three regions of electron density. Draw the molecules and measure their bond angles. How do nonbonding (lone pair—LP) electrons explain the molecular geometry of SO_2? Explain any differences you notice between the theoretical bond angles (see the theoretical models) and your observed values.

C. Examine the molecules with four regions of electron density. Draw the molecules and measure their bond angles. How do nonbonding (lone pair—LP) electrons explain the molecular geometries of NH_3 and H_2O? Explain any differences you notice between the theoretical bond angles (see the theoretical models) and your observed values.

D. Examine the molecules with five regions of electron density. Draw the molecules and measure their bond angles. How do nonbonding (lone pair—LP) electrons explain the molecular geometry of SF_4 and ClF_3? Explain any differences you notice between the theoretical bond angles (see the theoretical models) and your observed values.

E. Examine the molecules with six regions of electron density. Draw the molecules and measure their bond angles. How do nonbonding (lone pair—LP) electrons explain the molecular geometries of XeF_4 and IF_5? Explain any differences you notice between the theoretical bond angles (see the theoretical models) and your observed values.

II. Trends

A. Draw Lewis Structures for ONF, ONCl, ONBr, and ONI. Predict their molecular geometries and bond angles. Record this information in the following space.

B. Examine these molecules with a molecular viewer. Record the actual bond angles and bond lengths in the following space. Account for any differences between your predictions in part A. with your findings. Account for any trends you notice in bond angles and bond lengths.

III. Conformation

A. Using your molecular viewer, examine the C_4H_{10} molecule. Orient the molecule so that all of the atoms are visible and draw the structure in the following space.

B. Configure the molecule so that you can rotate the molecule around the bond between the second and third carbon atoms. Orient the molecule so that this bond is perpendicular to the plane of the computer screen. Rotate the bond until the two end carbon atoms are as far apart as possible. Reorient the molecule until you can see all of the atoms and draw the structure in the following space. Compare your two drawings of C_4H_{10}. Which orientation (conformation) would you predict would be the most stable? Why?

IV. Isomers

If two molecules have the same molecular formulas but different bonding arrangements for the atoms, the molecules are said to be isomers of each other. There are several kinds of isomers. A good test for the presence of isomers is to try to superimpose the molecules on each other. If molecules can be superimposed, they are identical. Sometimes it is necessary to rotate the atoms around single bonds (free rotation is not possible around double or triple bonds) to test the superimposable nature of molecules. In this exercise you will be given sets of molecules with the same molecular formulas. You should view these sets, manipulate them in an attempt to superimpose the molecules, identify any identical molecules, identify isomers, and characterize how the isomers are structurally different.

A. View the molecules of your first set. Identify their molecular formula, draw the structures of the molecules, and state your conclusions.

B. View the molecules of your second set. Identify their molecular formula, draw the structures of the molecules, and state your conclusions.

C. View the molecules of your third set. Identify their molecular formula, draw the structures of the molecules, and state your conclusions.

D. View the molecules of your fourth set. Identify their molecular formula, draw the structures of the molecules, and state your conclusions.

V. Carbon Compounds

A. Allotropes

These files contain three ways that elemental carbon is found in nature. View each of these "allotropic forms" of carbon and characterize each.

B. Bond Types

These files contain structures of molecules containing two carbons connected by single, double, and triple bonds. Draw Lewis Structures of these molecules and indicate their carbon-to-carbon bond lengths. Determine the carbon-to-carbon bond lengths for benzene (C_6H_6). How would you characterize the nature of the carbon-to-carbon bond for benzene?

C. Cyclic Compounds

Compare and contrast the three six-carbon cyclic structures in these files.

VI. Constructing a Molecule

A. Open the CO_2 molecule file with your computer viewing program. View the molecule in the "ball and stick" format. See your instructor if you don't know how to do this. Draw a picture of what you see on your screen.

B. Open the same CO_2 molecule file with a text reading program. Copy the text of the file in the following space. Identify the following information contained in the file: (1) the location of each of the three atoms in the molecule, (2) how the atoms are connected to each other, and (3) the length of the bonds connecting each oxygen to carbon.

C. Change the text file and see how the change affects the appearance of the molecule in the viewing program. Indicate what change you made and draw the resulting molecule in the following space.

D. Draw a Lewis Structure of SF_6 and indicate its molecular geometry. Modify or construct a text file to represent the SF_6 molecule. Assume a SF bond length of 1.79 Å. Copy the text file below.

PROPERTIES OF CALCIUM IODATE—Exp. I-2

Name________________________________ Lab Section ________________________

Lab Partner________________________

Pre-Lab Assignment: Study Appendix D.1 on volumetric measurements and titrations.

Problem Statement: What controls the solubility of $Ca(IO_3)_2$?

I. Preparation of Solutions

To save time one partner should work on part A below while the other partner works on part B.

NOTE: There are two stock solutions of both $Ca(NO_3)_2$ and KIO_3 available. Be sure to check the concentrations to insure that you are using the proper solution.

A. Preparations of $Ca(IO_3)_2(s)$

Obtain 20 mL of 23 g/100 mL $Ca(NO_3)_2$ solution and 70 mL of 5 g/100 mL KIO_3 solution in separate clean beakers. Pour the solutions together and stir. Rub the stirring rod against the sides or bottom of the beaker. (This "scratching" the solution is used to speed the formation of the precipitate.)

The precipitate formed is $Ca(IO_3)_2$. Let this solution stand for 10 minutes and then filter (See Appendix D.5). Using the supernatant, wash any precipitate that remains in the beaker onto the filter paper until about 90% of the solid has been trapped on the paper. Wash the trapped precipitate with four separate 10 mL portions of distilled water.

B. Preparation of $Ca(NO_3)_2$ Solutions

Obtain 90 mL of 0.1 M $Ca(NO_3)_2$ in a clean, dry beaker. Record its exact concentration in the space below. 40 mL of this solution will be used in part C; the remainder will be used to make less concentrated solutions of about 0.025 M and 0.006 M.

Rinse a 25 mL pipet with a 5 mL portion of the 0.1 M solution. Using a pipetting bulb (See Appendix D.1) transfer 25.00 mL of the 0.1 M solution into a 100.00 mL volumetric flask. Fill to the mark with distilled water and mix by inversion. Label a clean beaker and rinse it with two 5 mL portions of the solution. Store the solution in the beaker.

Rinse the pipet and volumetric flask with distilled water and then prepare a 0.006 M solution in a similar manner using the 0.025 M solution. Calculate the concentrations of these two prepared solutions to 3 significant figures in the following space.

C. Label four clean 100 mL beakers as #1 to #4. Put 40 mL of a different solution (distilled water, 0.006 M Ca^{2+}, 0.025 M Ca^{2+}, 0.1 M Ca^{2+}) into different beakers. Each beaker should be rinsed with a small portion of its solution before the 40 mL is added. Record which beaker contains which solution in Table 2 on page 161.

Put about one-quarter of the wet $Ca(IO_3)_2$ precipitate into each beaker. Stir each mixture (use a separate stirring rod for each to prevent contamination). Let the solutions stand for about 30 minutes with occasional stirring. Record any observations below. In the meantime proceed with part II.

II. Data Collection and Analysis: Standardization of $Na_2S_2O_3$ Solution

A. Obtain 100 mL of sodium thiosulfate ($Na_2S_2O_3$) solution in a clean, labeled beaker. Rinse a buret with small portions of the solution and fill the buret with the solution.

B. Obtain about 50 mL of 0.01 M KIO_3 solution, 10 cm^3 of solid KI, 20 mL of 1 M HCl and 20 mL of 0.2% starch solution. Obtain a medicine dropper for the HCl solution and one for the starch solution. Using a bulb, rinse the 10 mL pipet with three 2 mL portions of the 0.01 M KIO_3. Pipet 10.00 mL of this solution into each of three distilled-water rinsed 250 mL Erlenmeyer flasks. Add 10 to 20 mL of distilled water. Dissolve about 1 cm^3 of solid KI into each flask and add two droppersful of 1 M HCl (approximately 2 mL). Swirl to mix.

The coloring of the solution comes from the I_2 formed by the oxidation-reduction reaction:

$$IO_3^-(aq) + 5\,I^-(aq) + 6H^+(aq) \rightarrow 3I_2(aq) + 3H_2O$$

C. Record the initial buret volume reading of thiosulfate solution in Table 1 on page 159. Select one I_2 solution flask and titrate it with the $S_2O_3^{2-}$ solution (See Appendix D.1 for pointers). Be sure to swirl the flask after each ***small*** addition of $S_2O_3^{2-}$. The end-point of the titration is reached at just that point where the titrated solution is permanently colorless. It is at this point that the colored I_2 has just been consumed by the $S_2O_3^{2-}$. The reaction is:

$$2\,S_2O_3^{2-}(aq) + I_2\,(aq) \rightarrow S_4O_6^{2-}(aq) + 2I^-(aq)$$

The color of the solution will fade from yellow-brown to pale yellow as the titration progresses. When the pale yellow color is reached, add a dropperful of starch solution. The starch simply enhances the coloring of I_2, making it easier to determine the end point. Continue the titration until the colorless end-point is reached. Record the final buret reading and the total volume of $S_2O_3^{2-}$ solution used.

Titrate the other I_2 solutions. If the results of the three titrations do not agree to within 1%, run a fourth titration.

D. Calculate the molar concentration of the stock $S_2O_3^{2-}$ solution from the known concentration of your IO_3^- solution and the combination reaction (See B and C above):

$$IO_3^-(aq) + 6\ H^+(aq) + 6\ S_2O_3^{2-}(aq) \rightarrow I^-(aq) + 3\ S_4O_6^{2-}(aq) + 3\ H_2O$$

Show your calculations for the second titration in the space below. Calculate an average value for $S_2O_3^{2-}$ solution concentration.

TABLE 1

Standardization of Stock Thiosulfate Solution

Concentration of KIO_3 solution titrated ____________________ M

Titration No.	1	2	3	4
mL KIO_3	________	________	________	________
initial buret reading (mL)	________	________	________	________
final buret reading (mL)	________	________	________	________
total $S_2O_3^{2-}$ volume (mL)	________	________	________	________
molarity of $S_2O_3^{2-}$ (M)	________	________	________	________

average $S_2O_3^{2-}$ molarity ____________M

III. Data Collection: The Solubility of $Ca(IO_3)_2$

A. You are to filter each solution prepared in part I.C so that a 20 to 40 mL sample of uncontaminated and undiluted supernatant is obtained. Label four clean and dry beakers. Prepare a dry filter paper cone for each of four clean and dry filter funnels. Pass each solution through a different dry paper cone, catching the supernatant in the appropriately labeled beaker. The solid $Ca(IO_3)_2$ and filter papers can be discarded.

B. Label four distilled-water rinsed Erlenmeyer flasks and pipet 10.00 mL of each supernatant into a different flask.

Add 10–20 mL of distilled water, 1 cm^3 of solid KI and two droppersful of 1 M HCl to each flask. Record your observations.

C. What conclusions can be drawn from these observations? (Hint: What ions must be present for the colored I_2 to form? Where did each of these come from in these experiments? Considering the data collected to this point, what can be said about the extent to which $Ca(IO_3)_2$ dissolves?)

D. Titrate each of these solutions as was done in part II. Record the titration data in Table 2. If time permits, titrate a second 10.00 mL sample of each supernatant.

Using the molar concentration of stock thiosulfate solution determined in part II and the analysis reaction given there, determine the molar concentration of IO_3^- developed in each supernatant solution. Record these values in Table 2. Show your calculations on supernatant #2 in the space below.

TABLE 2

Analysis of Working Solutions

Solution No.	1	2	3	4
initial [Ca^{2+}], (M)	______	______	______	______
initial buret reading (mL)	______	______	______	______
final buret reading (mL)	______	______	______	______
total $S_2O_3^{2-}$ volume (mL)	______	______	______	______
molarity IO_3^- (M)	______	______	______	______

E. Transfer from Table 2 to Table 3 the initial concentration of Ca^{2+} and the final concentration of IO_3^- for each supernatant. Assuming that $Ca(IO_3)_2$ completely breaks into IO_3^- and Ca^{2+} ions when it dissolves, calculate the concentration of $Ca(IO_3)_2$ that dissolved and the final concentration of Ca^{2+}. Show the calculations for the second supernatant below.

TABLE 3

Supernatant Compositions (all concentrations in M)

Solution No.	1	2	3	4
initial [Ca^{2+}]	______	______	______	______
final [IO_3^-]	______	______	______	______
[$Ca(IO_3)_2$] dissolved	______	______	______	______
final [Ca^{2+}]	______	______	______	______

IV. Data Analysis

A. Mental Model—Looking at the data in Table 3, identify a qualitative relationship between the amount of IO_3^- dissolved and the amount of Ca^{2+} present. Draw a picture showing this effect.

B. Find an algebraic expression that best relates the final concentrations of IO_3^- and Ca^{2+} to each other. (Hint: Try all possible mathematical combinations of the concentrations for a given supernatant, using multiplication and division and squaring and cubing the concentrations. See which combination gives the most constant result for all supernatants.) What practical uses could your algebraic expression have?

IRON (III) NITRATE AND POTASSIUM THIOCYANATE
Exp. I-5

Name________________________________ Lab Section ____________________________

Lab Partner____________________________

Pre-Lab Assignment: Review the textbook discussion of molar (mole/liter) concentration units. Study Appendix D.1 on buret volume measurements and Appendix D.2 on the theory and use of spectrophotometers.

Problem Statement: What controls the amount of reaction between Fe^{3+} and SCN^{-}?

I. Data Collection and Analysis: Qualitative

A. Dissolve about 1 cm^3 of hydrated iron (III) nitrate, $Fe(NO_3)_3(H_2O)_9$, in 20 mL of distilled water. Dissolve about 1 cm^3 of potassium thiocyanate, KSCN, in another 20 mL of water. Label each solution and describe the appearance of each.

B. Mix about 10 mL of each solution in a third beaker. Describe the results. (Set aside the remaining 10 mL of the $Fe(NO_3)_3$ and KSCN solutions for use in part III.C).

C. The observed reaction is: $Fe^{3+}(aq) + SCN^{-}(aq) \rightarrow FeSCN^{2+}(aq)$. Identify the color of each reactant and product ion.

II. Data Collection—Quantitative: $Fe^{3+}(aq) + SCN^{-}(aq) \rightarrow FeSCN^{2+}(aq)$

You will be using a spectrophotometer to measure the color intensity and thus the concentration of the $FeSCN^{2+}(aq)$ formed in several solutions. Since the experiments depend on concentrations, it is important for the precision of the results that the volume measurements be made with care. Review Appendix D.1 on the proper use of burets. Also, you should be careful not to unwittingly dilute or contaminate your solutions.

A. Label eight dry 150 mm test tubes as numbers 1 to 8. Obtain in separate, dry, labeled beakers 80 mL of KSCN solution and 80 mL of $Fe(NO_3)_3$ solution. Record the exact molar concentration of each solution in the table on page 179. (Each solution contains HNO_3 at 0.3 M to prevent the formation of iron hydroxides.)

Clean two burets. Rinse one with two 5 mL portions of the KSCN, and rinse the other with $Fe(NO_3)_3$ solution. Fill each buret with the appropriate solution, making sure the tip is free of air gaps. Rinse and fill a third buret with distilled water. Label each buret.

B. You are to prepare eight solutions of the two reactants such that each contains the same amount of KSCN but differing amounts of $Fe(NO_3)_3$ in 20.00 total mL. This will be done by carefully buretting amounts of the stock solutions into the test tubes. The amounts suggested are given in the following table.

Test Tube	mL Stock KSCN	mL Stock $Fe(NO_3)_3$	mL H_2O
1	5.00	0.50	14.50
2	5.00	1.00	14.00
3	5.00	1.50	13.50
4	5.00	2.00	13.00
5	5.00	2.50	12.50
6	5.00	10.00	5.00
7	5.00	13.00	2.00
8	5.00	15.00	0.00

Make up your solutions recording the exact volumes used (to ± 0.01 mL) in the table on page 179. The order of addition of the reagents is not important.

Mix each solution by stoppering with a rubber stopper and inverting the tube several times. Shaking will not properly mix the reagents. Be sure to wipe the stopper dry between use with different test tubes.

C. Read the color intensity (absorbance) of the $FeSCN^{2+}$ in each solution and record it in the table. In using the spectrophotometer follow the directions provided by the instructor and/or those given in Appendix D.2. The light wavelength of 447 nm is suggested for the measurements.

Retain at least 5 mL of test solution #2 for use in part III.C. If a spectrophotometer is not immediately available for your use, do parts III.A, B, and C while you are waiting.

Data Table

Molarity of stock KSCN solution ______________________M

Molarity of stock $Fe(NO_3)_3$ solution ______________________M

Test Tube	mL Stock KSCN	mL Stock $Fe(NO_3)_3$	mL H_2O	Total added SCN^- conc (M)	Total added Fe^{3+} conc (M)	Absorbance at 447 nm
1	________	________	________	________	________	________
2	________	________	________	________	________	________
3	________	________	________	________	________	________
4	________	________	________	________	________	________
5	________	________	________	________	________	________
6	________	________	________	________	________	________
7	________	________	________	________	________	________
8	________	________	________	________	________	________

III. Data Analysis

A. Calculate the total M concentration of SCN^- and Fe^{3+} added to each test solution. Enter these in the table on page 179. Show the calculations for test solution #5 below.

B. Considering that fact that Fe^{3+} and SCN^- react 1:1 to form the colored ion $FeSCN^{2+}$, what is the ***maximum*** mole/liter concentration of $FeSCN^{2+}$ that could form in each test solution? Explain your reasoning below. (Hint: What is the limiting reagent in each solution?)

C. Into each of 3 small test tubes put 10 drops (½ mL) of test solution #2. Into one put 5 drops of the KSCN solution you prepared in part I.A. Into the second test tube put 5 drops of the $Fe(NO_3)_3$ solution from part I.A; into the third put 5 drops of H_2O. Mix the solutions and compare them. Describe the results. What do the results indicate about the completeness of the reaction in test solution #2? Briefly explain your reasoning.

D. What conclusions can be drawn from the data on page 179 about the completeness of the reaction in each test solution? Offer an explanation for your findings. It may be helpful to graph the absorbance vs. Fe^{3+} data. (Hints: Since Absorbance = constant x [$FeSCN^{2+}$], how would the absorbances compare if the reaction went to completion in each solution? How can the actual absorbances be explained? Why might the reaction apparently go to completion in some situations but not in others?)

IV. Mathematical Analysis

A. To analyze the data mathematically, we need to determine the constant in the equation:

$$Abs = constant \times [FeSCN^{2+}]$$

This can be done by inspecting the graph in part III.D. If this graph has a plateau, we can assume these solutions are about 100% reacted. Thus their [$FeSCN^{2+}$] can be calculated.

Using this approach, determine a value for the constant in the equation above. Then use this value with the non-plateau solutions and compute [$FeSCN^{2+}$], [SCN^-], and [Fe^{3+}]. Enter the data in the following table. Show one of these calculations in the space below.

Test Tube	Actual Concentrations (M) of			Results: Possible Combinations		
	$FeSCN^{2+}$	SCN^-	Fe^{3+}			
1						
2						
3						
4						
5						
6						
7						
8						

B. What conclusions can be drawn from these data? Find an algebraic equation that relates the actual concentrations of $FeSCN^{2+}$, SCN^-, and Fe^{3+} to each other in the non-plateau solutions. (Hint: Try all possible combinations of the three concentrations by multiplication and/or division. For example, multiply all three together, multiply two and divide by the third, etc., looking for the combination that gives the most constant result.) Summarize your results in the table above. Be sure to label each column. Discuss your results below.

OXIDATION-REDUCTION REACTIONS AND VOLTAIC CELLS—Exp. J-1

Name______________________________ Lab Section ______________________________

Lab Partner______________________________

Problem Statement: How do metals and metal ions react?

I. Data Collection: Interaction of Metals and Metal Ions

A. Clean and rinse with distilled water, four 12 x 75 mm test tubes (small test tubes). Shake excess water out of the tubes. Fill each of the tubes half full of $Zn(NO_3)_2$ solution. Clean strips of Zn, Cu, Pb, and Ag metals by gently sandpapering the last one inch of each strip. Submerge the cleaned end of the strips into the solutions for about 30 seconds. Record your observations in the following table.

B. Rinse the four test tubes and clean the metal strips. Fill the tubes half full of $Cu(NO_3)_2$ solution and place the metal strips in the solutions. Record your observations.

C. Repeat the above procedure with $AgNO_3$ solution and then with $Pb(NO_3)_2$ solution.

	Zn(s)	Cu(s)	Pb(s)	Ag(s)
$Zn^{2+}(aq)$				
$Cu^{2+}(aq)$				
$Ag^{+}(aq)$				
$Pb^{2+}(aq)$				

II. Data Analysis

In the boxes of the table, write balanced chemical equations for each reaction observed. (If you observed no change in the system write "no reaction.") Pick two of these reactions and explain why you chose the products you did.

III. Interpretation

A. What patterns are shown in these data? (Hint: Compare the relative reactivity of the metals. Compare the relative reactivity of the metal ions. Identify any connections between these comparisons.)

B. Mental Model—Pick one of the chemical reactions from part I and construct a picture that shows how metals and metal ions interact. Explain in words how your picture illustrates your observations.

IV. Data Collection: Voltaic Cells

A. Fill a 100 mL beaker 1/3 full of 1 M NH_4NO_3. Obtain a U-tube with a length of cord inside it and invert this system into the solution.

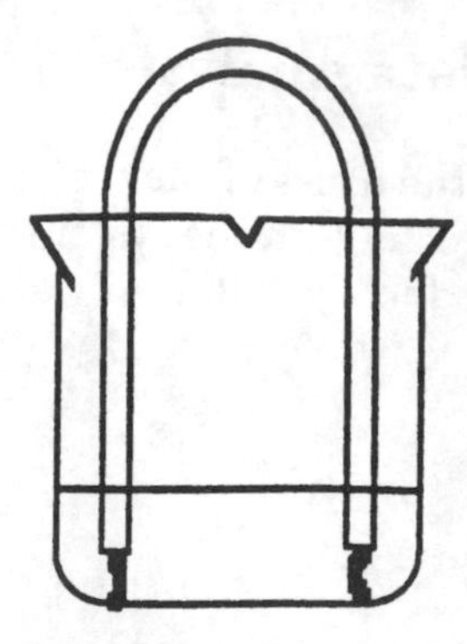

B. Clean, rinse with distilled water, and shake dry four 12 x 75 mm test tubes. Label the test tubes Zn, Pb, Cu, and Ag. Fill the test tubes ½ full of $Zn(NO_3)_2$, $Pb(NO_3)_2$, $Cu(NO_3)_2$, and $AgNO_3$ solutions, respectively. Put a cleaned strip of Zn metal in the $Zn(NO_3)_2$ solution, a cleaned strip of Pb in the $Pb(NO_3)_2$ solution, a cleaned Cu wire in the $Cu(NO_3)_2$ solution, and a cleaned strip of Ag in the $AgNO_3$ solution. Bend the end of the metal strips over the edges of the test tubes. Store the test tubes in a 100 mL beaker.

C. Assemble a Zn/Pb cell. Do this by placing the Zn test tube (called the Zn half cell and indicated by the symbol $Zn/Zn^{2+}||$) and Pb test tube (called the Pb half cell and indicated by the symbol $Pb/Pb^{2+}||$) into a 100 mL beaker. Remove the U-tube from the NH_4NO_3, blot the strings on a paper towel, and connect the two test tubes by immersing one leg of the U-tube in each of the two test tubes.

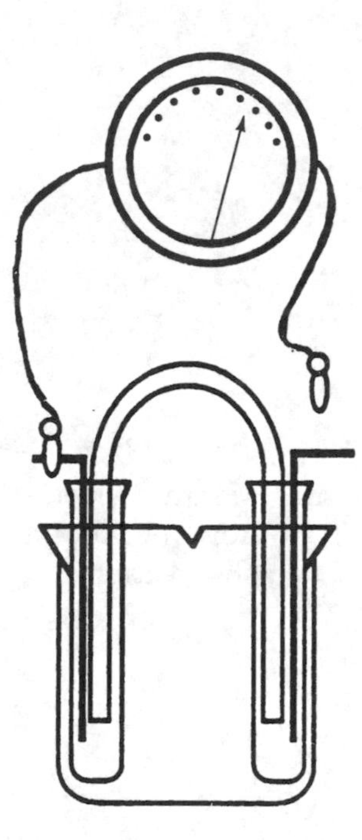

D. Set a volt meter to measure 0-2 volts. Clip one lead to the Zn metal strip and the other to the Pb strip. Switch the meter on or connect the meter just long enough to take a stable reading. Then turn the meter off (or disconnect it).

E. Assemble each of the possible six cells from the four half cells, and record the voltage of each cell. Rinse (by dipping the ends of the U-tube in water and then in NH_4NO_3) and blot the U-tube between readings.

F. Go through the entire procedure again, obtaining a second reading for each combination of half cells. Record these readings and compute an average reading for each combination of half cells.

Cell = ½ cell (1) + ½ cell (2)	Voltage Readings First	Second	Average
1. Zn/Zn^{2+} || Pb^{2+}/Pb			
2. Zn/Zn^{2+} || Cu^{2+}/Cu			
3. Zn/Zn^{2+} || Ag^{+}/Ag			
4. Ag/Ag^{+} || Pb^{2+}/Pb			
5. Ag/Ag^{+} || Cu^{2+}/Cu			
6. Cu/Cu^{2+} || Pb^{2+}/Pb			

V. Data Analysis

Electricity is usually viewed as a flow of electrons. Explain why these experimental set-ups cause an electrical current to flow through the wires and volt meter. (Hint: Choose one combination of cells and suggest what reactions may be occurring in each half cell. How do the reactions of the two half cells combine to give a net flow of electrons? What might be the net reaction of the combined half cells?)

VI. Interpretation

A. What patterns exist in the voltage readings for the several combinations of half cells? How do these patterns compare to those found in part III.A?

B. What relationships are there among the voltages of the $Zn/Zn^{2+} || Pb^{2+}/Pb$, $Zn/Zn^{2+} || Cu^{2+}/Cu$, and $Pb/Pb^{2+} || Cu^{2+}/Cu$ cells? Explore the relationships of other similarly related cells.

C. Mental Model—Pick one of the voltaic cells you constructed in part IV and draw an illustration of how the electrons and ions (both positive and negative) move through the cell.

VII. Data Collection: Concentration Effects

A. The Cu^{2+}/Cu half cell will be used in this experiment. You will also need Ag^+ solutions of the following molarities: 0.20 M (solution from the previous experiment), 0.020 M, 0.0020 M, and 0.00020 M. These last three can be prepared by a technique called serial dilution. The 0.020 M solution is prepared by diluting 1.0 mL of 0.20 M Ag^+ with 9.0 mL of distilled water. The 0.0020 M solution is prepared by diluting 1.0 mL of the 0.020 M Ag^+ with 9.0 mL of distilled water. In a similar manner the 0.00020 M Ag^+ solution is prepared from the 0.0020 M Ag^+ solution.

Place a few mLs of each Ag^+ solution in separate, clean, dried, labeled 12 x 75 mm test tubes.

B. Place the cleaned Ag strip in the 0.00020 M Ag^+ solution and connect this half cell to the 0.2 M Cu^{2+}/Cu half cell. Measure the voltage as described in part IV. Record the data in the following table.

Remove the Ag metal strip and salt bridge, rinse them both and construct a cell using the 0.0020 M Ag^+ solution. Measure its voltage. Repeat this procedure with the 0.020 M and 0.20 M Ag^+ solutions.

C. Obtain a second reading of the voltage for each cell. Record these in the table and compute an average reading for each cell.

	Voltage Readings		
Cell	**First**	**Second**	**Average**
1. Ag/0.00020 M Ag^+ || Cu^{2+}/Cu	______	______	______
2. Ag/0.0020 M Ag^+ || Cu^{2+}/Cu	______	______	______
3. Ag/0.020 M Ag^+ || Cu^{2+}/Cu	______	______	______
4. Ag/0.20 M Ag^+ || Cu^{2+}/Cu	______	______	______

VIII. Data Analysis and Interpretation

What conclusion can be drawn from these data? Can the pattern of the data be expressed as a simple algebraic equation? Justify your answers. It might be helpful to graph the data.

ELECTROLYSIS REACTIONS—Exp. J-2

Name____ Lab Section ________________________________

Lab Partner______________________________

Problem Statement: How is electrical energy absorbed in a chemical reaction?

I. Data Collection and Analysis: Lead and Copper Reactions

A. Obtain approximately 50 mL of 0.2M solutions of $Cu(NO_3)_2$ and $Pb(NO_3)_2$ and place each in separate 100 mL beakers. Clean a copper and lead electrode by sanding lightly one inch of the bottom tip. Place the lead electrode in the $Cu(NO_3)_2$ solution and the copper electrode in the $Pb(NO_3)_2$ solution. Let them remain in solution for 5 minutes. Record any observations.

Write chemical equations for any changes you noticed.

B. Place a second Cu electrode in the $Pb(NO_3)_2$ solution separated from the first Cu electrode. Attach a D.C. current source to the electrodes. Note which electrode is attached to which pole of the source. Follow your instructor's instructions about how to use your D.C. current source. Your apparatus should look like the diagram. Switch on the power and allow it to run for approximately 5 minutes.

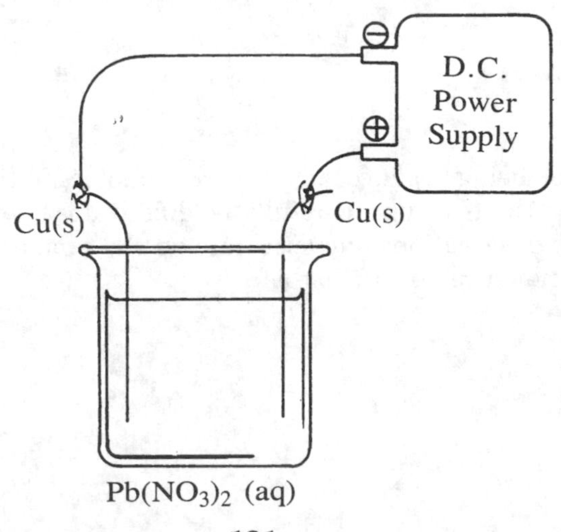

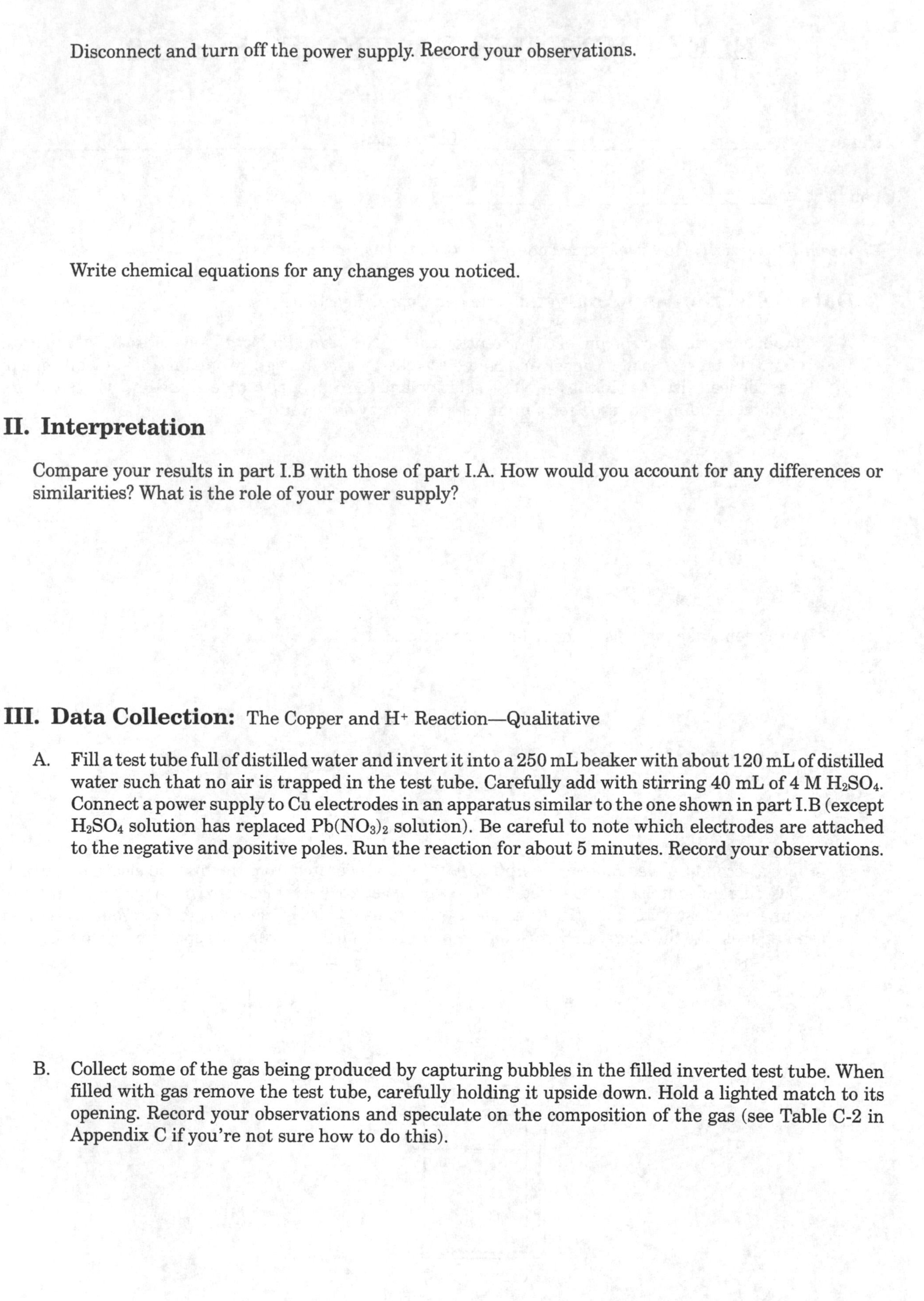

Disconnect and turn off the power supply. Record your observations.

Write chemical equations for any changes you noticed.

II. Interpretation

Compare your results in part I.B with those of part I.A. How would you account for any differences or similarities? What is the role of your power supply?

III. Data Collection: The Copper and H^+ Reaction—Qualitative

A. Fill a test tube full of distilled water and invert it into a 250 mL beaker with about 120 mL of distilled water such that no air is trapped in the test tube. Carefully add with stirring 40 mL of 4 M H_2SO_4. Connect a power supply to Cu electrodes in an apparatus similar to the one shown in part I.B (except H_2SO_4 solution has replaced $Pb(NO_3)_2$ solution). Be careful to note which electrodes are attached to the negative and positive poles. Run the reaction for about 5 minutes. Record your observations.

B. Collect some of the gas being produced by capturing bubbles in the filled inverted test tube. When filled with gas remove the test tube, carefully holding it upside down. Hold a lighted match to its opening. Record your observations and speculate on the composition of the gas (see Table C-2 in Appendix C if you're not sure how to do this).

IV. Data Analysis

A. Write a half-reaction describing the chemistry at the electrode connected to the negative pole of the power supply. (Use table C-6 in Appendix C for possible half-reactions. The half-reactions in Table C-6 may be reversed if that will better describe the reaction being studied.)

B. Write a half-reaction describing the chemistry at the electrode connected to the positive pole of the power supply.

V. Interpretation

Combine the two half-reactions in parts A and B to describe the overall reaction occurring in the beaker. Draw a diagram of the apparatus tracing the flow of electricity.

VI. Data Collection: The Copper and H^+ Reaction—Quantitative

A. Obtain and clean a 50 mL buret. Using a 10 mL graduate, measure the volume contained by the buret between the 50 mL mark and the stopcock. Record this value on the data table on page 195.

Put about 375 mL of distilled water in a 600 mL beaker. Carefully add 125 mL of 4 M H_2SO_4 and stir. The solution will be 1 M. (Your instructor may have an alternative method for obtaining this solution.)

Invert the buret into the solution and carefully fill it by opening the stopcock and drawing the solution into the buret using a suction bulb. Close the stopcock when the space below the stopcock is just filled. Leave the tip empty. Check the system for leaks.

Obtain a length of copper wire with the ends stripped and the middle covered with water-tight plastic. Bend and place the wire so that one stripped end is entirely inside the buret and the other end is outside the beaker (see diagram).

B. Clean a Cu electrode by sanding, dipping it into 6M nitric acid, rinsing it with water, and drying it. Weigh the electrode as accurately as is possible with the balances available to you. Record the value on the data table.

Bend the electrode and place it in the solution. Connect the leads from the Cu electrode to the + pole of the power source/ammeter and the –pole to the copper wire (see diagram). Note the time to the nearest second and quickly adjust the current to between 0.10 amp and 0.15 amp according to the ammeter. Record the time and current on the data table. Check the current every 10 minutes and adjust if necessary.

Obtain an atmospheric pressure and temperature reading and record these on the data table.

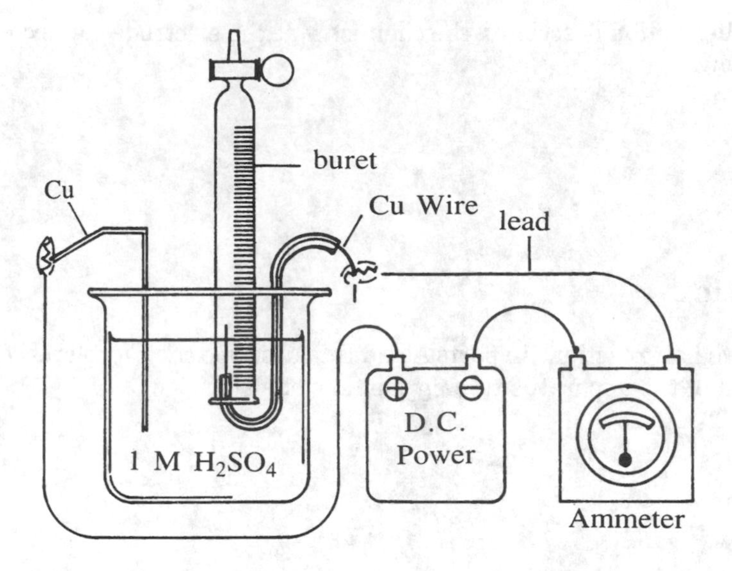

C. Six times during the progress of the reaction, record on the data table the volume reading and the elapsed time. Be careful when reading the graduations on the buret to note they are upside down. Determine the actual volume of gas collected and record this on the table.

D. Sometime before the collected gas reaches the zero mark on the buret, turn off or disconnect the power, noting the exact time to the second. Also note the elapsed time and the final volume reading. Record these data on the data table.

E. Very carefully remove the Cu electrode from the solution. Dip the electrode into a beaker of distilled water, being careful not to dislodge any material. Dip the electrode into acetone and then allow it to air dry until no more moisture remains. Weigh the electrode and record this data on the data table.

Data Table

Volume of buret between 50 mL ⟶ stopcock ________mL

Initial mass Cu ______________________g   Final mass Cu ______________ g

Time began __________________________   Time finished ________________

Current ________________________ amps

Atmospheric Pressure _____________ torr

Temperature ______________________ °C

	Volume Reading	Time Elapsed (sec)	Volume Gas Collected (mL)
		0	0
final reading			

VII. Data Analysis

A. What is the relationship between the volume of gas and the elapsed time? Develop an algebraic equation which expresses this relationship. It might be helpful to graph your data and develop the algebraic equation from the graph.

B. Why are volume of gas and time related in the manner shown in part IV.A? (Hint: How is the current related to production of gas according to the half-reactions developed in part III? How is this related to your algebraic equation in section VII.A?)

C. Calculate the number of moles of gas produced during the experiment. Show your work. (Don't forget to adjust the partial pressure of your gas sample to allow for the presence of water vapor.) The vapor pressure of water at various temperatures can be found in the *Handbook of Chemistry and Physics*.

D. From your weights of Cu, calculate the number of moles of Cu consumed during the experiment. Show your work.

E. Compare the amount of H_2 and Cu produced or consumed during the experiment. How does this relate to the amount of electrons utilized? How does this relate to your proposed chemical equation from part V?

VIII. Interpretation

A. Write a generalization which summarizes your findings in this section of the experiment. (Hint: How much silver would be produced (or consumed) if it was used instead of Cu in the cell? See the half-reactions in Table C-6 in Appendix C.)

B. (Optional) The quantity of electricity (coulombs) = current (amps) x time (sec). Calculate the relationship between the quantity of electricity and amount of electrons in an electrochemical reaction.

KINETIC FACTORS—Exp. K-1

Name____________________________________ Lab Section __________________________________

Lab Partner________________________________

Problem Statements: What factors can be used to change the rate of a reaction?
Are these factors predictable from the balanced equation?

I. The Reaction: Permanganate Ion Reduced by Oxalic Acid

In this experiment you will be studying the reaction between permanganate, MnO_4^{-1}, and oxalic acid, $C_2O_4H_2$, in acid solution. The overall balanced reaction equation is:

$$2\ MnO_4^{-1}(aq) + 5\ C_2O_4H_2(aq) + 6\ H^{+}(aq) \rightarrow 2\ Mn^{+2}(aq) + 10\ CO_2(aq) + 8\ H_2O$$

II. Watching the Reaction

A. Using clean, dry, labeled 100 mL beakers, obtain 30 to 50 mL each of the stock solutions 0.0014 M K_2MnO_4, 0.20 M $C_2O_4H_2$ and 2.0 M H_2SO_4. Put about 50 mL of distilled water in a fourth beaker. Put a clean dropper in each beaker. You will be making up various mixtures of these four solutions by counting drops. Be careful that the droppers are not switched between beakers during any of the work.

B. Take three clean, distilled-water rinsed, large test tubes and label them 1 through 3. You will be doing experiments three runs at a time and need to keep track of each run. These can be supported in a test tube rack or in a 250 mL beaker while you observe them. Viewing them against a white background is advised.

C. Make up the three solutions as described below. Mix each by flicking the bottom of the tube while holding its top securely. Observe these solutions for about 10 minutes and record your observations.

Drops of Stock Solutions

Run #	$KMnO_4$	H_2O	H_2SO_4	$C_2O_4H_2$
1A	15	15	15	0
2A	15	25	0	5
3A	15	10	15	5

D. Observations

E. Data Analysis—Offer explanations for your observations. Did all runs show a change? If not, why not? Did those that changed do so instantly? If not, why not?

III. Data Collection: Effect of H^+ Concentration

A. In the experimental runs from here on, you will be timing how long it takes for all the MnO_4^{-1} ions to react by watching the color fade. The recommended procedure is as follows: Dump the current contents of your test tubes into your waste beaker and rinse each a few times with about 1 mL of water. Shake out excess water. Add the stock solutions to each tube as called for ***except*** for the $C_2O_4H_2$. Then quickly add the drops of $C_2O_4H_2$ to each tube, mix by flicking, and record this as the starting time. Also record the time when the color has faded.

	Drops of Stock Solutions				Times		
Run #	$KMnO_4$	H_2O	H_2SO_4	$C_2O_4H_2$	Starting	Ending	Total
1B	15	20	5	5	______	______	______
2B	15	10	15	5	______	______	______
3B	15	0	25	5	______	______	______

B. What effect does increasing the concentration of H^+ ion have on the rate of the reaction? Does this seem reasonable? Explain your reasoning in terms of what the molecules and ions may be doing.

IV. Data Collection and Analysis: Effect of $C_2O_4H_2$ Concentration

A. Make up the runs below using the procedure outlined in part II above.

	Drops of Stock Solutions				Times		
Run #	$KMnO_4$	H_2O	H_2SO_4	$C_2O_4H_2$	Starting	Ending	Total
1C	15	10	15	5	______	______	______
2C	15	5	15	10	______	______	______
3C	15	0	15	15	______	______	______

B. What effect does increasing the concentration of oxalic acid have on the rate of the reaction? Does this seem reasonable? Explain your reasoning in terms of what the molecules and ions may be doing.

V. Data Collection and Analysis: Effect of Copper Metal

A. Do the runs as outlined below using the procedure in part II. Again put everything together except the $C_2O_4H_2$, then start counting time with its addition. Cut copper metal turnings in lengths of about 2 and 4 inches. Bunch each into a very loose ball that can be submerged in the solution and still have all of the turning exposed to solution.

	Drops of Stock Solutions					Times		
Run #	$KMnO_4$	H_2O	H_2SO_4	$C_2O_4H_2$	Cu Length	Starting	Ending	Total
1D	15	10	15	5	0	______	______	______
2D	15	10	15	5	2 in.	______	______	______
3D	15	10	15	5	4 in.	______	______	______

B. What effect does the presence and amount of Cu metal have on the rate of the reaction? Does this seem reasonable? Explain your reasoning in terms of what the molecules and ions may be doing.

VI. Data Collection and Analysis: Effect of Temperature

A. Using a ring stand, ring, wire gauze and burner, make a hot water bath in a half-filled 250 mL beaker. Heat the bath until it is 40–50 °C. Maintain this temperature interval by heating or allowing the bath to cool.

Make up the runs below ***except*** for the $C_2O_4H_2$. Put the test tubes into the bath for several minutes to get the solution to temperature. Remove one run from the bath, quickly add the 5 drops of $C_2O_4H_2$, mix, and return to the bath. When that reaction is done, repeat the experiment on the other run.

	Drops of Stock Solutions					Times		
Run #	$KMnO_4$	H_2O	H_2SO_4	$C_2O_4H_2$	Temp	Starting	Ending	Total
1E	15	10	15	5	40–50 °C	______	______	______
2E	15	10	15	5	40–50 °C	______	______	______

B. What effect does increasing the temperature have on the rate of the reaction? Does this seem reasonable? Explain your reasoning in terms of what the molecules and ions may be doing.

VII. Conclusions

A. What could a chemist do to this redox reaction to speed it up or slow it down? Which factors are more effective? Which are less effective? Explain your reasoning.

B. Could some or all the factors you found to affect the reaction rate have been predicted from knowing only the overall reaction? Explain your reasoning.

C. Mental Model—Select one of the four effects you investigated in this experiment and draw pictures showing why you think it works as it does. That is, look at your explanations in part B of these experiments, select one and turn the verbal explanation into a pictorial explanation of how molecules and ions react. Label the effect for clarity.

QUALITATIVE ANALYSIS FOR Ag^+, Pb^{2+}, Ni^{2+}, Ba^{2+}, Fe^{3+}—Exp. L-1

Name_________________________________ Lab Section _____________________________

Lab Partner___________________________

Pre-Lab Assignment: Study Appendix D.4 on qualitative analysis techniques.

Problem Statement: How can we determine if the cations Ag^+, Pb^{2+}, Ni^{2+}, Ba^{2+}, and Fe^{3+} are present in a solution if one, two, or all might be there?

PART ONE

I. Data Collection

A. Obtain approximately 10 mL of 0.1 M solutions of Ag^+, Pb^{2+}, Ni^{2+}, Ba^{2+}, and Fe^{3+}in separate clean, labeled beakers. (Each of these solutions was prepared from the nitrate salt of the cation.) You are to subject each of these solutions to the six tests outlined below and record your observations in the table provided.

B. Chemical Tests:

1. Put 10 drops of test solution into a 75 mm test tube and then add 5 drops of 3 M H_2SO_4.

2. To 10 drops of fresh test solution add 5 drops of 6 M HCl. If a precipitate forms, mix the sample and centrifuge. Wash the precipitate twice with H_2O and save it for test 3.

3. Add 1½ mL of fresh distilled water to the precipitate from test 2; mix. Heat the tube in a boiling water bath for 4 minutes. Mix frequently. Centrifuge and decant the hot liquid. To the hot liquid add 1 drop of K_2CrO_4.

4. To 10 drops of fresh test solution add 10 drops of 6 M NaOH. If no precipitate develops, tests 5 and 6 need not be done. If a precipitate forms, mix the sample and centrifuge. Discard the supernatant and save the precipitate for test 5.

5. Wash the precipitate from test 4 once with H_2O. Add 10 drops of 15 M NH_3 to the precipitate and mix for one minute. Centrifuge.

6. Repeat test 4 to obtain a hydroxide precipitate free of supernatant. Wash the precipitate once with H_2O. Add 5 drops of 6 M HCl to the precipitate and mix. Add 2 drops of 1 M NH_4SCN solution.

C. Observations

Test and Reagent	CATION				
	Ag^+	Pb^{2+}	Ni^{2+}	Ba^{2+}	Fe^{3+}
1. SO_4^{2-}					
2. Cl^-					
3. a. heat					
b. CrO_4^{2-}					
4. OH^-					
5. NH_3					
6. a. H^+					
b. SCN^-					

II. Data Analysis: Possible Equations

From your data complete the equations listed below. You will be speculating, but your speculation will be accurate in 90% of the cases if you use the following guidelines: (1) Assume that most reactions are a simple combination of cations and anions. (2) If the product is dissolved it is ionic. (3) If the product is solid it is neutral. If you observed no change when the particular reagents were mixed, write N.R. for "no reaction."

1. $Ag^{+} + SO_4^{2-} \longrightarrow$ ______

2. $Pb^{2+} + SO_4^{2-} \longrightarrow$ ______

3. $Ni^{2+} + SO_4^{2-} \longrightarrow$ ______

4. $Ba^{2+} + SO_4^{2-} \longrightarrow$ ______

5. $Fe^{3+} + SO_4^{2-} \longrightarrow$ ______

6. $Ag^{+} + Cl^{-} \longrightarrow$ ______ $\xrightarrow{+\text{ heat}}$ ______ $\xrightarrow{+CrO_4^{2-}}$ ______

7. $Pb^{2+} + Cl^{-} \longrightarrow$ ______ $\xrightarrow{+\text{ heat}}$ ______ $\xrightarrow{+CrO_4^{2-}}$ ______

8. $Ni^{2+} + Cl^{-} \longrightarrow$ ______ $\xrightarrow{+\text{ heat}}$ ______ $\xrightarrow{+CrO_4^{2-}}$ ______

9. $Ba^{2+} + Cl^{-} \longrightarrow$ ______ $\xrightarrow{+\text{ heat}}$ ______ $\xrightarrow{+CrO_4^{2-}}$ ______

10. $Fe^{3+} + Cl^{-} \longrightarrow$ ______ $\xrightarrow{+\text{ heat}}$ ______ $\xrightarrow{+CrO_4^{2-}}$ ______

11. $Ag^{+} + OH^{-} \longrightarrow$ ______ $\xrightarrow{+NH_3}$ ______

    $Ag^{+} + OH^{-} \longrightarrow$ ______ $\xrightarrow{+H^{+}}$ ______ $\xrightarrow{+SCN^{-}}$ ______

12. $Pb^{2+} + OH^{-} \longrightarrow$ ______ $\xrightarrow{+NH_3}$ ______

    $Pb^{2+} + OH^{-} \longrightarrow$ ______ $\xrightarrow{+H^{+}}$ ______ $\xrightarrow{+SCN^{-}}$ ______

13. $Ni^{2+} + OH^{-} \longrightarrow$ ______ $\xrightarrow{+NH_3}$ ______

    $Ni^{2+} + OH^{-} \longrightarrow$ ______ $\xrightarrow{+H^{+}}$ ______ $\xrightarrow{+SCN^{-}}$ ______

14. $Ba^{2+} + OH^{-} \longrightarrow$ ______ $\xrightarrow{+NH_3}$ ______

    $Ba^{2+} + OH^{-} \longrightarrow$ ______ $\xrightarrow{+H^{+}}$ ______ $\xrightarrow{+SCN^{-}}$ ______

15. $Fe^{3+} + OH^{-} \longrightarrow$ ______ $\xrightarrow{+NH_3}$ ______

    $Fe^{3+} + OH^{-} \longrightarrow$ ______ $\xrightarrow{+H^{+}}$ ______ $\xrightarrow{+SCN^{-}}$ ______

III. Data Analysis: An Experimental Plan

A. Assume you were given a solution that was said to contain Ba^{2+} cations only. Describe how you would confirm or deny the presence of this cation in the solution. Explain your reasoning.

B. Assume you were given a solution that was said to contain Pb^{2+} and Ag^{+} cations only. Describe how you would determine if neither, only one, or both of these ions were present in solution. Explain your reasoning.

C. Assume you were given a solution that was said to contain Fe^{3+} and Ni^{2+} cations. Describe how you would determine if neither, only one, or both of these ions were present in solution. Explain your reasoning.

D. Assume you were given a solution that was said to contain Ag^{+}, Pb^{2+}, Ni^{2+}, Fe^{3+}, and Ba^{2+} cations only. Describe how you would confirm or deny the presence of each of these ions in the solution. Explain your reasoning. (Hint: You may wish to design an analysis scheme—a flow chart—such as the one shown in the Appendix on page 337.)

PART TWO

Analysis of an Unknown

You will be given 10 mL of a solution of unknown composition which might contain any or all of the cations: Ag^{+}, Pb^{2+}, Fe^{3+}, Ba^{2+}, and Ni^{2+}. You are to analyze your solution in whatever manner you feel is appropriate to determine unequivocally which of the cations are present.

The lab report for this section should include the following:

1. ***An experimental section***. This should be a step-by-step accounting of the experiments conducted and the results observed. Include balanced ionic equations where appropriate.

2. ***A conclusion section***. This should be a paragraph telling which cations you think were in your solution and telling why you reached this conclusion.

QUALITATIVE ANALYSIS FOR Zn^{2+}, Ca^{2+}, Cu^{2+}, Al^{3+}, Co^{2+}
Exp. L-2

Name______________________________________ Lab Section ________________________________

Lab Partner________________________________

Pre-Lab Assignment: Study Appendix D.4 on qualitative analysis techniques.

Problem Statement: How can we determine if the cations Zn^{2+}, Ca^{2+}, Cu^{2+}, Al^{3+}, Co^{2+} are present in a solution if one, two, or all might be there?

PART ONE

I. Data Collection

A. Obtain approximately 10 mL of 0.1 M solutions of Zn^{2+}, Ca^{2+}, Cu^{2+}, Al^{3+}, and Co^{2+} in separate, clean, labeled beakers. (Each of these solutions was prepared from the nitrate salt of the cation.) You are to subject each of these solutions to the five tests outlined below and record your observations in the table provided.

B. Chemical Tests

1. Put 10 drops of test solution into a 75 mm test tube. To this neutral solution, add 7 drops of 0.3 M Na_3PO_4.

2. To 10 drops of fresh test solution add 4 drops of 3 M HNO_3 to make the solution strongly acidic, $pH<2$. Add 7 drops of 0.2 M $K_3Fe(CN)_6$, potassium ferricyanide.

3. To 10 drops of fresh test solution add 10 drops of 4 M NaOH. If a precipitate forms, mix the sample and centrifuge. Decant the supernatant and save the precipitate for test 4.

4. Wash the precipitate from test 3 three times with distilled water. Make each wash about 1 mL of water. To the final washed ppt. add 10 drops of 15 M NH_3 and mix for 1 minute.

5. Repeat step 3 to obtain a hydroxide precipitate free of supernatant. Wash the precipitate twice with distilled water. Add 15 drops of 3 M HNO_3 to the precipitate and mix.

C. Observations

CATION

Test and Reagent	Zn^{2+}	Ca^{2+}	Cu^{2+}	Al^{3+}	Co^{2+}
1. PO_4^{3-}					
2. $Fe(CN)_6^{3-}$					
3. OH^-					
4. NH_3					
5. H+					

II. Data Analysis: Possible Equations

From your data complete the equations listed below. You will be speculating, but your speculation will be accurate in 90% of the cases if you use the following guidelines: (1) Assume that most reactions are a simple combination of cations and anions. (2) If the product is dissolved it is ionic. (3) If the product is solid it is neutral. If you observed no change when the particular reagents were mixed, write N.R. for "no reaction."

1. $Zn^{2+} + PO_4^{3-} \longrightarrow$ ________
2. $Ca^{2+} + PO_4^{3-} \longrightarrow$ ________
3. $Cu^{2+} + PO_4^{3-} \longrightarrow$ ________
4. $Al^{3+} + PO_4^{3-} \longrightarrow$ ________
5. $Co^{2+} + PO_4^{3-} \longrightarrow$ ________

6. $Zn^{2+} + Fe(CN)_6^{3-} \longrightarrow$ ________
7. $Ca^{2+} + Fe(CN)_6^{3-} \longrightarrow$ ________
8. $Cu^{2+} + Fe(CN)_6^{3-} \longrightarrow$ ________
9. $Al^{3+} + Fe(CN)_6^{3-} \longrightarrow$ ________
10. $Co^{2+} + Fe(CN)_6^{3-} \longrightarrow$ ________

11. $Zn^{2+} + OH^- \longrightarrow$ ________ $\xrightarrow{+NH_3}$ ________

    $Zn^{2+} + OH^- \longrightarrow$ ________ $\xrightarrow{H^+}$ ________

12. $Ca^{2+} + OH^- \longrightarrow$ ________ $\xrightarrow{+NH_3}$ ________

    $Ca^{2+} + OH^- \longrightarrow$ ________ $\xrightarrow{+H^+}$ ________

13. $Cu^{2+} + OH^- \longrightarrow$ ________ $\xrightarrow{+NH_3}$ ________

    $Cu^{2+} + OH^- \longrightarrow$ ________ $\xrightarrow{+H^+}$ ________

14. $Al^{3+} + OH^- \longrightarrow$ ________ $\xrightarrow{+NH_3}$ ________

    $Al^{3+} + OH^- \longrightarrow$ ________ $\xrightarrow{+H^+}$ ________

15. $Co^{2+} + OH^- \longrightarrow$ ________ $\xrightarrow{+NH_3}$ ________

    $Co^{2+} + OH^- \longrightarrow$ ________ $\xrightarrow{+H^+}$ ________

III. Data Analysis: An Experimental Plan

A. Assume you were given a solution that was said to contain Ca^{2+} cations only. Describe how you would confirm or deny the presence of this cation in the solution. Explain your reasoning.

B. Assume you were given a solution that was said to contain Zn^{2+} and Al^{3+} cations only. Describe how you would determine if neither, only one, or both of these ions were present in solution. Explain your reasoning.

C. Assume you were given a solution that was said to contain Cu^{2+} and Co^{2+} cations only. Describe how you would determine if neither, only one, or both of these ions were present in solution. Explain your reasoning.

D. Assume you were given a solution that was said to contain Zn^{2+}, Ca^{2+}, Cu^{2+}, Al^{3+}, and Co^{2+} cations only. Describe how you would confirm or deny the presence of each of these ions in the solution. Explain your reasoning. (Hint: You may wish to design an analysis scheme—a flow chart—such as the one shown in the Appendix on page 337.)

PART TWO

Analysis of an Unknown

You will be given 10 mL of a solution of unknown composition which might contain any or all of the anions: Cl^-, I^-, PO_4^{3-}, $C_2O_4^{2-}$, and SO_4^{2-}. You are to analyze your solution in whatever manner you feel is appropriate to determine unequivocally which of the anions are present.

The lab report for this section should include the following:

1. ***An experimental section***. This should be a step-by-step accounting of the experiments conducted and the results observed. Include balanced ionic equations where appropriate.

2. ***A conclusion section***. This should be a paragraph telling which anions you think were in your solution and telling why you reached this conclusion.

53 & 54

IRON ANALYSIS—Exp. M-1

Name______________________________ Lab Section ___________________________

Lab Partner__________________________

Pre-Lab Assignment: Study Appendix D-1 on titration techniques.

Problem Statement: How can we determine the amount of Fe in a sample?

PART ONE

I. Data Collection and Analysis: Qualitative Tests for Fe^{2+} and Fe^{3+}

A. Obtain about 20 mL of 0.1 M $FeSO_4$ (Fe^{2+} or ferrous ions) and 20 mL of 0.1 M $Fe(NO_3)_3$ (Fe^{3+} or ferric ions) in distilled-water rinsed, labeled beakers. Also obtain about 20 mL of 2 M H_2SO_4, 5 mL of 0.03 M $K_2Cr_2O_7$, 5 mL of 1 M NH_4SCN, and 5 mL of 0.02 M $KMnO_4$ in clean, labeled beakers.

B. Place about 5 mL of water, 5 mL of Fe^{2+} solution and 5 mL of Fe^{3+} solution in separate rinsed, labeled, large test tubes or small beakers. Record the appearance of each in the table below. Add 1 mL of 2 M H_2SO_4 (about one dropperful) to each and record their appearance. Finally, add 4 drops of 0.03 M $K_2Cr_2O_7$ to each, mix, and record their appearance.

C. Using 1 mL of H_2SO_4, acidify fresh 5 mL samples of H_2O, Fe^{2+}, and Fe^{3+} solution. Add 1 drop of 1 M NH_4SCN to each solution and record your observations.

	Water blank	**Fe^{2+} ion**	**Fe^{3+} ion**
No Addition			
acid			
$Cr_2O_7^{2-}$			
SCN^-			

D. Explain briefly how $Cr_2O_7^{2-}$ and SCN^- can be used to test any solution for the presence of Fe^{2+} and/or Fe^{3+}.

II. Data Collection and Analysis: Qualitative—Reaction of Iron Ions with MnO_4^-

A. **NOTE**: when MnO_4^- decolors, it is being converted to Mn^{2+}.

Acidify fresh 5 mL samples of H_2O, Fe^{2+}, and Fe^{3+} with 1 mL portions of H_2SO_4. Add 15 to 20 drops of 0.02 M MnO_4^- to each and mix. Record your observations. Save the solutions.

Water blank:

Fe^{2+} solution:

Fe^{3+} solution:

B. If any of these solutions are colorless, we can test them for the type of iron ions they now contain. Split any colorless iron ion solution into two parts. Add 4 drops of $Cr_2O_7^{2-}$ to one part and 1 drop of SCN^- to the other. Record your observations.

C. Do Fe^{2+} ions react with MnO_4^-? If you feel a reaction occurs, identify the products of the reaction. Explain in detail how your data support your conclusions.

Write a balanced chemical equation for the reaction of Fe^{2+} with MnO_4^-.

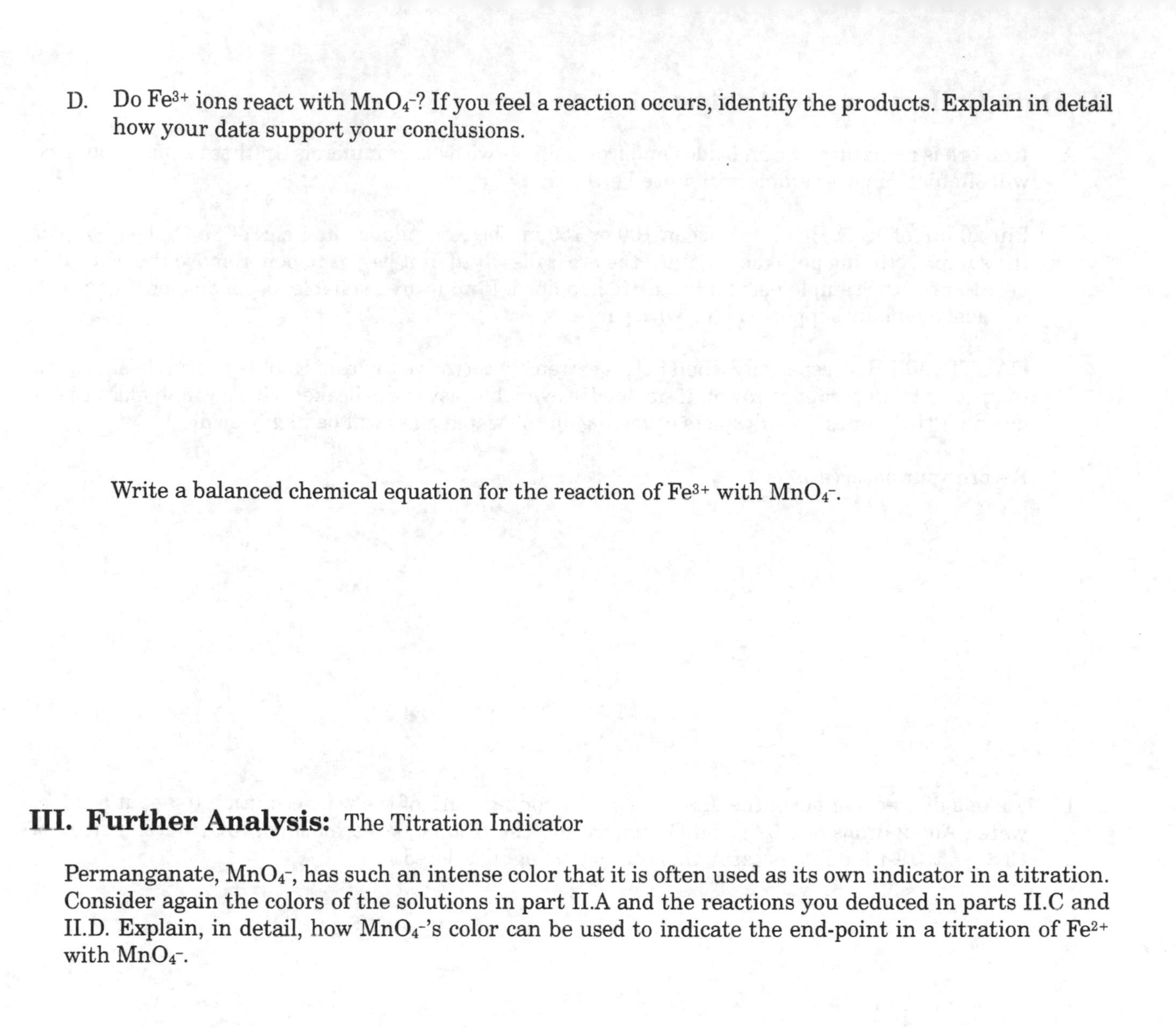

D. Do Fe^{3+} ions react with MnO_4^-? If you feel a reaction occurs, identify the products. Explain in detail how your data support your conclusions.

Write a balanced chemical equation for the reaction of Fe^{3+} with MnO_4^-.

III. Further Analysis: The Titration Indicator

Permanganate, MnO_4^-, has such an intense color that it is often used as its own indicator in a titration. Consider again the colors of the solutions in part II.A and the reactions you deduced in parts II.C and II.D. Explain, in detail, how MnO_4^-'s color can be used to indicate the end-point in a titration of Fe^{2+} with MnO_4^-.

IV. Data Collection and Analysis: Iron Ore

A. Iron ore is a mixture of iron oxides and iron sulfides with other minerals. In these experiments we will simulate an ore sample with pure Fe_2O_3.

Put 20 mL of 12 M HCl into a clean 100 or 150 mL beaker. Add about 1 cm^3 of Fe_2O_3. Gently heat the sample, stirring periodically, until the ore is dissolved. If it begins to boil, remove the heat. It is best to heat the sample using a hot plate in a hood. If none are available, use a burner flame with the beaker solidly supported on a wire gauze.

CAUTION!!! HCl, especially when hot, is extremely corrosive to flesh. Handle the hot beaker with tongs or a triply wrapped towel. If no hood is available, cover the beaker with a watch glass to cut down on HCl fumes. The droplets collecting on the watch glass will be highly acidic.

Record your observations.

B. Make a diluted sample of the dissolved ore by adding 1 mL of the warm solution to about 5 mL of water. Add 2 drops of 0.02 M MnO_4^- and record the results. (We are using MnO_4^- here, instead of $Cr_2O_7^{2-}$, to test for Fe^{2+} because this solution is highly colored.)

Make a second diluted sample of the dissolved ore and add 1 drop of 1 M SCN^-. Record your observations.

What forms of iron ions are present in the dissolved ore? Explain your reasoning.

C. Obtain about 10 mL of 1 M $SnCl_2$. Add this Sn^{2+} a dropperful at a time to the remainder of the dissolved ore sample. Continue until the sample is colorless. The sample should be kept warm during these additions and should be constantly stirred.

Make two 1 mL:5 mL dilutions of the colorless ore solution (see part IV.B). Add 20 drops of $Cr_2O_7^{2-}$ to one dilution, one drop of SCN^- to the other. Record your observations.

What forms of iron ions are present in the ore solution after Sn^{2+} treatment? Write a balanced chemical equation for the reaction of the ionic iron with Sn^{2+} (assume the tin product is Sn^{4+}). Briefly explain your reasoning.

D. Describe how you would prepare a solid ore sample for analysis of total iron content by MnO_4^- titration. Explain why each step in the preparation is necessary.

V. Data Collection and Analysis: Titration of Iron Ion with MnO_4^-

A. Obtain about 50 mL of 0.02 M $KMnO_4$ solution in a clean, labeled beaker. Take precautions to not dilute or contaminate your sample of this solution. Prepare a buret and fill it with $KMnO_4$ solution as directed in Appendix D.1. Record the exact concentration of this solution in the table on the next page.

B. Weigh 0.5 to 0.8 grams of $FeSO_4(NH_4)_2SO_4(H_2O)_6$, ferrous ammonium sulfate, into a 250 mL Erlenmeyer flask. The exact mass should be known to at least three significant figures.

Add 100 mL of distilled water, 10 mL of 3 M HCl, then swirl to dissolve the Fe^{2+} salt.

(If the sample was an Fe^{3+} salt, you would now heat it to near boiling, add a couple of drops of 1 M $SnCl_2$ and swirl. You would continue this dropwise addition and mixing until the yellow color of Fe^{3+} was just dispelled.)

C. Add 10 mL of 3 M H_3PO_4 solution. (The phosphate sharpens the end point by binding to the iron ions.)

D. Record the initial volume reading of the buret, then titrate the Fe^{2+} sample. The end-point is reached when the addition of one drop of MnO_4^- causes a pink color that persists in the solution for about 30 seconds. Record the final buret reading and the total volume of MnO_4^- used.

E. If time allows, weigh out another sample of the Fe^{2+} salt and titrate it. Collect titration data from other students.

F. Complete the table by filling in the last two rows. Show your calculations for one set of data below.

Data Table

$KMnO_4$ solution = ____________________ M

Titration Data

grams of Fe^{2+} salt					
initial buret reading					
final buret reading					
mL MnO_4^- solution					
moles of Fe^{2+}					
moles of MnO_4^-					

VI. Conclusions

A. In part II.C you wrote a balanced chemical equation for the reaction of Fe^{2+} and MnO_4^-. Show mathematically how your titration data confirm or deny this equation. Explain your reasoning. If the equation does not fit the titration data, find a new equation that does fit both the qualitative data and titration data. Explicitly show that this new equation does fit all the data.

B. Mental Model—Draw a picture showing how you see Fe^{2+} ions interacting with MnO_4^- ions in solution. Be sure your picture shows the stoichiometry you have determined for that reaction.

PART TWO

Analysis of an Unknown

You are to apply the chemistry and techniques from Part One to determine the iron content of "unknown" samples. Your instructor will decide the basic objective of the analysis. The following are possibilities:

Alternative no. 1: You will be given a solution, salt or powdered ore unknown. You are to determine the iron concentration of it in mol/L or weight %. You will be judged on the aptness of your procedure, the reproducibility of your results, and the accuracy of your results. The purpose is to perfect your analysis skills.

Alternative no. 2: You will be given, or are to collect, several natural substances containing iron (raw ore samples, mine tailings, commercial products.) You are to analyze these for iron content, noting any analytical difficulties and any trends among related samples. You will be judged on the aptness of your procedure, the reproducibility of your results, and the explanations you develop. The purpose is to sharpen your analytical skills and to give you experience with "real world" analysis problems.

Your report on this work is to contain the following sequential parts:

1. A brief statement of the scope of the experiment
2. A brief outline of procedure
3. Presentations of raw data
4. Mathematical treatment of the data
5. Conclusions and explanations

WATER ANALYSIS—Exp. M-2

Name________________________________ Lab Section ________________________________

Lab Partner________________________________

Pre-Lab Assignment: Study Appendix D.1 on the use of burets and D.3 on the use of pH meters.

Problem Statements: What is water hardness, and how can we measure it?
What is water alkalinity, and how can we measure it?

PART ONE

I. Data Collection and Analysis: Qualitative on Hardness

A. Obtain approximately 5 mL of 0.01 M solutions of Ca^{2+}, Mg^{2+}, Na^{+}, K^{+}, and Fe^{3+} in separate, distilled-water rinsed, labeled beakers. (Each of these solutions was prepared from the nitrate salt of the cation.)

Obtain approximately 15 mL of soap solution in a clean, labeled beaker. Clean and distilled-water rinse 5 large test tubes.

B. Label the test tubes as Ca^{2+}, Mg^{2+}, Na^{+}, K^{+}, and Fe^{3+}. Place about 2 mL of soap solution into each. Now add 7 to 10 drops ($^1/_3$ to $^1/_2$ mL) of the corresponding cation solution to each test tube. Do not contaminate your solutions. Use separate droppers for each solution.

Sealing the top of the test tube with your thumb, shake each vigorously for 15 seconds. Record your observations and comparisons below. Do not discard the solutions.

C. Considering these data, which of the cations cause water to be "hard water"? Explain your answer. (i.e., How do you define "hard water"? Do any of your data fit this definition?)

D. Soap is a salt such as $CH_3(CH_2)_{16}CO_2^-Na^+$. Considering that the reaction involves the simple exchange of ions, write a balanced chemical equation for the reaction of soap with "hard water" cations.

E. List at least two instances in which the use of water with a high hardness level would be disadvantageous. List one instance in which it might be advantageous. Briefly explain your reasoning in each case.

II. Data Collection: Quantitative on Hardness

A. Obtain about 40 mL of the $Ca(NO_3)_2$ stock solution in a clean, labeled beaker. Record the exact concentration of the solution in the data table on the next page.

B. Obtain about 50 mL of the stock "EDTA" solution in a labeled beaker, taking precautions not to dilute or contaminate your sample of this solution. Record the molar concentration of this solution in the table. The EDTA solution is the titrant in this experiment. Prepare a buret and fill it with EDTA solution as directed in Appendix D.1.

EDTA is an abbreviation for the anion: ethylenediaminetetraacetate. The anion, pictured below, contains C_2H_4 (ethylene), two nitrogens (diamine) and four $^-CH_2CO_2^-$ groups (tetraacetate). The anion reacts with cations by bonding through its two nitrogens and several of its oxygens. Thus it wraps itself around the cation by bonding to it at several points.

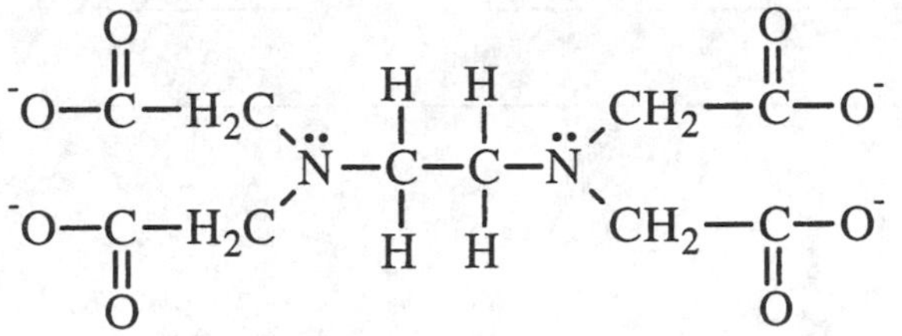

```
       O                                O
       ||          H  H                 ||
 ⁻O—C—H₂C         |  |        CH₂—C—O⁻
            \..    |  |   ../
             N—C—C—N
            /      |  |      \
 ⁻O—C—H₂C         H  H        CH₂—C—O⁻
       ||                               ||
       O                                O
```

C. Pipet exactly 10.00 mL of the stock $Ca(NO_3)_2$ solution into a distilled-water rinsed, 250 mL Erlenmeyer flask. Add distilled water to about the 50 mL mark. Obtain about 10 mL of stock NH_4Cl/NH_3 solution in a labeled, watch-glass covered beaker. Add 1 to 2 mL of this solution to the Erlenmeyer flask, using a clean medicine dropper.

(The NH_4Cl/NH_3 solution is a buffer that will keep the analysis solution at pH=10. This high pH prevents $EDTA^{4-}$ from turning into its conjugate acid, H_4EDTA.)

Add 2 drops of the indicator Eriochrome Black T (abbreviated EBT) to the flask. Record the initial buret reading in the table, then begin titrating the treated Ca^{2+} solution with $EDTA^{4-}$ solution. The titration is complete at just that point where the solution changes to a permanent ***pure*** blue color. Cautions to be observed are: (1) The color changes are a bit slow to develop. When near the end-point, swirl for 10 seconds between additions to allow any color changes to develop fully. (2) In some cases the color changes may progress from pink → purple → bluish purple → pure blue in the space of several drops. The end-point desired is pure blue, the first total absence of purple.

(The indicator, EBT, works by binding to certain cations until these all have been sucked up by the $EDTA^{4-}$.)

$M^{2+} \bullet EBT^{3-} \rightarrow EBT^{3-}$

(red) (blue)

D. Record your buret readings in the table. Repeat the titration with another 10.00 mL sample of the Ca^{2+} stock solution.

Hardness Data

Stock Solution Concentrations

$Ca(NO_3)_2$ solution = ______________________ M

EDTA solution = ______________________ M

		Buret Readings, in mL			Calculated Results	
Trial #	ml $Ca(NO_3)_2$	Initial	Final	Total	mol Ca^{+2}	mol EDTA
1	________	________	________	________	________	________
2	________	________	________	________	________	________
3	________	________	________	________	________	________

III. Data Analysis

A. Calculate the number of moles of Ca^{+2} ion and EDTA ion used in each trial from the volumes and concentrations of the stock solutions. Enter these numbers in the table above. Show your calculations for Trial #1 below.

B. Write a balanced chemical equation for the reaction of Ca^{+2} with $EDTA^{4-}$. Explain how your titration data support the balanced equation you've written.

IV. Data and Analysis—Qualitative: Other Cations and EDTA

A. Design and conduct quick, qualitative experiments to determine if K^+, Na^+, Mg^{2+}, or Fe^{3+} ions could be determined by reaction with $EDTA^{4-}$. Describe your experiments and results.

B. Which other cations besides Ca^{2+} can be determined by titration with $EDTA^{4-}$? How do these compare to the cations you found to cause hardness?

C. The total hardness of any natural water sample is reported in terms of the mg of $CaCO_3$ that would have to dissolve in one liter of water to give that same result. Consider the following data: "30.0 mL of water from river YY required 12.35 mL of 0.0114 M EDTA to titrate it to the end-point using the procedure in Section II."

Calculate the total hardness of river YY in units of mg/L of $CaCO_3$. Show your calculation and list any assumptions you had to make.

V. Data Collection: Quantitative on Alkalinity

A. Obtain a pH meter and buffer solution(s). Standardize the pH meter using the buffer(s) as directed by your instructor or as described in Appendix D.3.

B. Obtain about 70 mL of stock 0.02 M HCl solution in a clean, labeled beaker. Record the exact molarity of this solution in the table on the following page. The HCl solution is the titrant in this experiment. Prepare a buret and fill it with HCl solution as directed in Appendix D.1.

C. Using a graduated cylinder, measure out about 50 mL of the stock Na_2CO_3 solution. Record its exact concentration and volume in the table. Pour the 50 mL into a clean, distilled-water rinsed 100 or 150 mL beaker.

You are to eventually suspend the pH meter electrode into this solution in such a way that you also have the buret positioned above. The idea is to deliver measured amounts of HCl to the beaker without having to remove the electrode. You might play around with positioning and clamping using an empty 100 or 150 mL beaker.

D. Suspend the pH meter into the stock Na_2CO_3 solution. Position the HCl-filled buret above it. Measure the initial pH and the initial buret level; record these in the table. Take care to *not* allow any HCl to drip into the solution.

You are to add HCl in increments, stir, and record buret level and measured pH until a pH of 4.5 ± 0.1 is reached or passed. Do this in the following manner: Carefully add one or two drops of HCl solution to the Na_2CO_3 solution. Stir, measure the pH, and record the new buret reading. If the pH has fallen sharply (more than about 0.3 pH units), continue the titration one to two *drops* at a time. If the pH has not fallen much, continue the titration about 0.5 milliliters at a time.

E. Repeat steps C and D with stock $CaCl_2$, Na_3PO_4, Na_2SO_4 solutions and distilled water. Record your data in the table.

Alkalinity Data

stock HCl solution = ______________________ M

Na_2CO_3 solution		$CaCl_2$ solution		Na_3PO_4 solution		Na_2SO_4 solution		Dist. H_2O	
conc = _____ M		conc = _____ M		conc = _____ M		conc = _____ M			
vol = _____ mL		vol = _____ mL		vol = _____ mL		vol = _____ mL		vol = _____ mL	
buret reading	**pH**	**buret reading**	**pH**	**buret reading**	**pH**	**buret reading**	**pH**	**buret reading**	**pH**
_____	_____	_____	_____	_____	_____	_____	_____	_____	_____
_____	_____	_____	_____	_____	_____	_____	_____	_____	_____
_____	_____	_____	_____	_____	_____	_____	_____	_____	_____
_____	_____	_____	_____	_____	_____	_____	_____	_____	_____
_____	_____	_____	_____	_____	_____	_____	_____	_____	_____
_____	_____	_____	_____	_____	_____	_____	_____	_____	_____
_____	_____	_____	_____	_____	_____	_____	_____	_____	_____
_____	_____	_____	_____	_____	_____	_____	_____	_____	_____
_____	_____	_____	_____	_____	_____	_____	_____	_____	_____
_____	_____	_____	_____	_____	_____	_____	_____	_____	_____
_____	_____	_____	_____	_____	_____	_____	_____	_____	_____
_____	_____	_____	_____	_____	_____	_____	_____	_____	_____
_____	_____	_____	_____	_____	_____	_____	_____	_____	_____
_____	_____	_____	_____	_____	_____	_____	_____	_____	_____
_____	_____	_____	_____	_____	_____	_____	_____	_____	_____
_____	_____	_____	_____	_____	_____	_____	_____	_____	_____
_____	_____	_____	_____	_____	_____	_____	_____	_____	_____
_____	_____	_____	_____	_____	_____	_____	_____	_____	_____
_____	_____	_____	_____	_____	_____	_____	_____	_____	_____
_____	_____	_____	_____	_____	_____	_____	_____	_____	_____

VI. Data Analysis

A. Alkalinity is defined in *Water Analysis Manuals* as "the capacity of the water sample to accept protons." Which of the water samples you just analyzed do you feel have a high alkalinity? Which have little or no alkalinity? Explain your reasoning, justifying your decisions.

B. Mental Model—Draw a picture of what happens at the atom level when HCl is added to a water sample with a high alkalinity versus one with little or no alkalinity.

C. The total alkalinity of any natural water sample is measured as you did above, by measuring how much acid it takes to get that sample to a pH = 4.5. However, it is usually reported in terms of the mg of $CaCO_3$ that would have to dissolve in one liter of water to give that same result, assuming that each $CaCO_3$ molecule tied up only one H^+ ion.

Calculate the total alkalinity of the stock Na_2CO_3 solution from your data. Show that calculation below.

PART TWO

Analysis of an Unknown

You are to apply the chemistry and techniques from Part One to determine hardness and alkalinity in "unknown" samples. Your instructor will decide the basic objective of the analysis. The following are possibilities:

Alternative no. 1: You will be given a solid or solution unknown. You are to determine the hardness and alkalinity. You will be judged on the aptness of your procedure, the reproducibility of your results, and the accuracy of your results. The purpose is to have you perfect your analysis skills.

Alternative no. 2: You will be given, or are to collect, several related natural water samples. You are to determine the total hardness and alkalinity of each. You will be judged on the aptness of your procedure, the reproducibility of your results, and the patterns and explanations you develop. The purpose is to sharpen your analytical skills and to give you food for thought on the relationships between bodies of water.

Your report is to contain the following sequential parts:

1. Brief statement of the scope of the experiment
2. Brief outline of procedure
3. Presentation of raw data
4. Mathematical treatment of the data
5. Conclusions and explanations

DISSOLVED OXYGEN—Exp. M-3

Name______________________________________ Lab Section ________________________________

Lab Partner________________________________

Pre-Lab Assignment: Study Appendix D.1.E on use of burets.

Problem Statements: How can we analyze a water sample for the amount of O_2 dissolved in it?
How does dissolved O_2 relate to pollution levels?

PART ONE

I. Preparation of Solutions for Analysis

You are to prepare three solutions, using flasks specially designed to exclude air pockets. These will be called DO flasks (Dissolved Oxygen flasks). Your instructor will have demonstrated their use and safety precautions.

A. Secure three of these flasks, rinse them with distilled water, and label them A, B, and C. Also secure, in clean, labeled beakers: about 2g of d-glucose ($C_6H_{12}O_6$), and 15 mL of "microbe water"—a solution containing a high population of bacteria.

B. Fill each DO flask three-quarters full with distilled water. Stopper and shake the flask. Unstopper, restopper, and then shake again. This is meant to saturate the water with O_2 from the air. This is called aerating the sample.

You will need another 300 mL of aerated distilled water to top off your DO samples. Prepare this water using any appropriate flask.

C. Add the following to the designated DO flask, mix to dissolve:

Flask A: nothing more

Flask B: about 1g of d-glucose

Flask C: about 1g of d-glucose and 10 mL of microbe water

D. Fill each DO flask to its brim with aerated, distilled water. Stopper each over a sink. If air bubbles are trapped in the flask, take out the stopper and add more distilled water, then restopper. Mix each flask by inverting it several times.

Set the DO samples aside for later analysis in part III.

CAUTION!!! Both the KOH and the H_2SO_4 are extremely corrosive. If any gets on your person, wash the area immediately with water. Also, ***do not mix them at full strength***—the reaction is violent.

II. Data Collection and Analysis: Chemical Reactions

Listed below are chemical equations involved in the analysis of dissolved oxygen (DO) in a sample. Use these to help answer the questions in parts II.B to II.G.

$$I_2 + 2\ S_2O_3^{2-} \rightarrow S_4O_6^{2-} + 2\ I^-$$

$$2\ Mn^{2+} + 4\ OH^- + O_2 \rightarrow 2\ MnO_2(s) + 2\ H_2O$$

$$C_6H_{12}O_6 + 6\ O_2 \rightarrow 6\ CO_2 + 6\ H_2O$$

$$MnO_2(s) + 2\ I^- + 4\ H^+ \rightarrow Mn^{2+} + I_2 + 2\ H_2O$$

A. In separate, clean, towel-dried, labeled 100 mL beakers obtain about:

1 cm^3 of manganous chloride

3 pellets of potassium hydroxide

½ cm^3 of potassium iodide

1 cm^3 of sodium thiosulfate

20 mL conc. sulfuric acid (also save for part III)

B. Add 20 mL of water to the manganous chloride and mix. This will be called the reaction beaker. Record your observations below, speculate on what chemical reaction(s) occurred, and briefly explain your reasoning.

Observations:

Chemical Reaction(s):

Reasoning:

C. Add 20 mL of water to the KOH pellets, mix until dissolved. Add this solution to the reaction beaker, mix.

Observations:

Chemical Reaction(s):

Reasoning:

D. Dissolve the KI in 20 mL of water, add this to the reaction beaker, mix.

Observations:

Chemical Reaction(s):

Reasoning:

E. Using a dropper, add about ½ mL (10 drops) of conc. H_2SO_4 to the reaction beaker, mix.

Observations:

Chemical Reaction(s):

Reasoning:

F. Dissolve the $Na_2S_2O_3$ in about 20 mL of water. Add this a few milliliters at a time to the reaction beaker; mix between additions.

Observations:

Chemical Reaction(s):

Reasoning:

III. Conclusions

Use the information gained above to explain how you would quantitatively analyze a solution for its dissolved oxygen using a standardized solution of $Na_2S_2O_3$.

(You might describe what chemical treatment you would give the sample before titration, giving reasons for this treatment. Describe the titration. How will you know when you have reached the end-point? Chemically what happens at the end-point?)

IV. Data Collection: Titration of the DO Samples

A. Obtain about 50 mL of 0.02 M $Na_2S_2O_3$ solution in a labeled beaker. Take care not to dilute or contaminate your sample of this solution. Prepare a buret and fill it with the thiosulfate solution as directed in Appendix D.1.E. Record the exact concentration of the solution in the table on the next page.

B. Obtain, in clean, dry, labeled 100 mL beakers:

10 mL of 2 M Mn^{2+} solution
10 mL of 10 M KOH, 1 M KI solution
15 mL of conc. H_2SO_4 (from part II)
10 mL of starch solution

Dedicate a dropper to each solution. In doing the work below, ***take care not to mix up the droppers.***

C. Remove the stopper from DO flask A. Add 2 droppersful (about 1½ mL) of 2 M Mn^{2+} without touching the dropper to the flask's solution. Do this by placing the tip of the dropper just below the rim of the flask and letting the Mn^{2+} solution run down the inside of the rim.

Over a sink restopper the DO flask, taking care that no air bubbles are trapped inside. If air is trapped, remove the stopper, add distilled water, and restopper.

Mix the sample several times by inversion.

D. In like manner, add one dropperful of 10 M KOH, 1 M KI solution to the DO flask. Stopper over a sink, and mix by inversion. If any of the excluded solution gets on your hands, wash that area with copious amounts of water.

Let the solution sit for two minutes to allow the precipitate to settle.

E. Invert the DO flask to redisperse the precipitate. Remove the stopper and add 2 droppersful of concentrated H_2SO_4. Restopper and mix as before. Again, wash your hands if excluded solution gets on them.

Let the solution sit for a few minutes to clarify as much as possible.

F. Using a graduated cylinder, carefully measure 100 mL of DO solution A into a clean, distilled-water rinsed, 250 mL Erlenmeyer flask. Record the initial reading of the buret, then begin titrating the DO sample.

NOTE: The titration should only take a few milliliters of thiosulfate solution, so make small additions even at the beginning.

When the color of the solution has changed from a bright orange to a pale orange, add 2 droppers of starch solution. The remaining I_2 will complex with the starch, giving a deep blue color. Continue to titrate until the solution color changes to colorless or very pale blue. Record the final buret reading in the table.

G. Titrate DO samples B and C, using the procedure given in parts II.C to II.F above. Record these data in the table.

H. Collect another set of data from other students and enter these in the table.

Data Table

Concentration of $Na_2S_2O_3$ = ______________________ M

Titration Data

Sample:	A	B	C	A'	B'	C'
Contents:	________	________	________	________	________	________
	________	________	________	________	________	________
	________	________	________	________	________	________
mL sample titrated	________	________	________	________	________	________
initial buret reading	________	________	________	________	________	________
final buret reading	________	________	________	________	________	________
DO (mg O_2/L)	________	________	________	________	________	________

V. Data Analysis

A. Calculate the concentration of dissolved oxygen (DO) in milligrams of O_2 per liter for each sample. You will need to use the mole relationship discovered in part II.

Enter these numbers in the table. Show your calculation for your sample "A" below.

B. Water saturated with air at 25 °C has an O_2 concentration of about 8 mg/L. Compare your results to this number. Offer explanations for why your samples agree or disagree with this number.

C. A standard analysis is to put aerated water, a pollutant and bacteria in a DO flask and then analyze it 5 days later by titration as you did. The drop in DO over that time is called the Biochemical Oxygen Demand (BOD) of the pollutant. Explain why this is an appropriate label.

(You might speculate on what is happening chemically in the sample, and on what role the bacteria play.)

PART TWO

You are to apply the chemistry and techniques from Part One in experiments of your own design. Possible ideas are listed below. Because of time and equipment restraints, your instructor may organize investigative teams or limit your options. Write up your reports as described for on page 276 for other open inquiry experiments.

NOTE: The American Public Health Association, in "Standard Methods for the Analysis of Waters and Wastewaters," has established guidelines for BOD analysis. These use natural bacterial waters (i.e., lake waters) and require five or more days for complete reaction. To get the reactions to occur in the few hours of the student lab period, your instructor may have you use hyperactivated colonies of bacteria.

System 1

Investigate the DO level of various water samples. Examples might include fast-flowing streams; still, deep lakes; water from a nonaerated aquarium; water that has had N_2 bubbled through it.

System 2

Determine the experimental BOD of d-glucose and compare it to the theoretical value.

System 3

Investigate the effect of bacteria concentration on the rate of DO depletion.

System 4

Investigate the effect of nutrient (pollutant) concentration on the rate of DO depletion.

System 5

Compare the rates of DO depletion with bacteria from different sources. Possibilities include lake water, "grey water," bakers yeast, pre-incubated versus not incubated (hyperactivated versus natural).

ORGANIC FUNCTIONAL GROUPS—Exp. N-1

Name________________________________ Lab Section ________________________

Lab Partner__________________________

Problem Statement: What specific parts of organic molecules control their chemistry?

I. Data Collection: Qualitative Tests

A. You are to run the five tests below on each of eight organic compounds and record the results in the table on the next page. You are to do these tests using small test tubes or a microscale plate. Whatever the case, be sure to clearly label your test tubes and rack or your plate with the compound and test.

Obtain, from the instructor, a set of dispensing pipets/vials containing the eight organic compounds. Run a given test on all eight compounds at once, to better compare the results. If you are using test tubes, take these to the hood between tests to dispose of the waste. Then rinse each test tube 2 to 3 times with small amounts of hexane and shake them dry before using them in the next test. If you are using well plates, run as many tests as possible in the well plate provided and then dispose of waste as directed.

If you are counting drops to get milliliters, use 1 mL = 30 drops.

B. Chemical Tests

1. Cu^{2+}: Put 0.5 mL of a 10% $CuSO_4$ solution into the test tube or well. Now add 2 to 4 drops of the organic compound. If the compound is a solid (#G), add several grains of it to its copper sulfate solution. Mix and observe.

2. pH: Put 0.5 mL of distilled water into the test tube or well. Add 7 to 10 drops of compound and mix. If the compound is a solid, add an amount the size of a pea. Mix to try to dissolve. Test the aqueous solution with pH paper.

3. CO_3^{2-}: To the solutions created in test #2, add about 0.5 mL of 10% Na_2CO_3 solution.

4. Ce^{4+}: Put 7 to 10 drops of the compound into the test tube or well. If the compound is a solid, dissolve an amount ***half*** the size of a pea in 10 drops of hexane. Add 1 or 2 drops of 4% ceric ammonium nitrate solution.

5. Br_2: Put 7 to 10 drops of the compound into the test tube or well. If the compound is a solid, dissolve an amount ***half*** the size of a pea in 10 drops of hexane. Add 1 or 2 drops of 2% bromine (Br_2) solution.

C. Observations

Compound	Test Cu^{2+}	pH	CO_3^{2-}	Ce^{4+}	Br_2
A. C_2H_6O					
B. $C_2H_4O_2$					
C. $C_4H_{11}N$					
D. $C_5H_{12}O$					
E. $C_5H_{13}N$					
F. C_6H_{10}					
G. $C_7H_6O_3$					
H. C_8H_{18}					

II. Data Analysis

A. Patterns in the Data

From these data, determine which compounds have the same chemical properties (i.e., which show the same types of reactions or non-reactions). Group similar compounds together and briefly explain your reasoning.

B. Reactivity versus Molecular Structure

On page 267 are given the Lewis Structures of the eight organic compounds. Look at the atomic arrangements of the compounds in each "reacting group" you identified in question II.A above. Within each group, what specific parts of the organic molecules most likely control their chemistry? Briefly explain your reasoning.

III. The Reaction Between an Acid and an Alcohol

A. **Data Collection:** Running and Observing the Reaction

1. Using a ring stand, wire gauze and Bunsen burner, set up a hot water bath in a 150 or 250 mL beaker. The water level should be 1 to 2 inches deep. You will eventually want the bath to be at or just below boiling.

2. Into a clean, large test tube (6 inch) put the two organic compounds listed below. (If you are counting drops, take 1.0 mL to be 30 drops.) Mix them by flicking the bottom of the tube and carefully smell the result. Record the result below.

   1.5 mL of compound D, $C_5H_{12}O$, called iso-amyl alcohol

   1.0 mL of compound B, $C_2H_4O_2$, called acetic acid

3. Take the tube to a hood and slowly add 1.0 mL of 9 M H_2SO_4 by running drops down the inside of the test tube. Mix again by flicking the bottom of the tube while holding the top securely.

4. Stand the test tube up in your water bath. Let it heat without mixing for about 10 minutes. Record any changes you see within the tube's contents. Every few minutes, gently (no mixing) remove the tube and smell its vapor. Then return it to the bath. You can keep the bath at or just below boiling by bringing the flame in and out from below the bath beaker.

5. Record your observations. Are there indications that a reaction has occurred? Has the odor changed? If so, what fruit odor is it beginning to resemble?

B. **Data Analysis:** Identifying the Product

Look at the organic compounds and their properties listed in Appendix Table C-7. From your data, determine what compound you have formed. Briefly explain your reasoning.

C. **Interpretation**: A Reaction Mechanism

Using Lewis Structures and arrows, draw a picture of how compound D, the alcohol, probably reacted with compound B, the acid. What other product most likely formed?

Does this picture confirm or deny what you determined in part III.B about what parts of an organic molecule control its chemistry? Explain your reasoning.

Lewis Structures

A. C_2H_6O

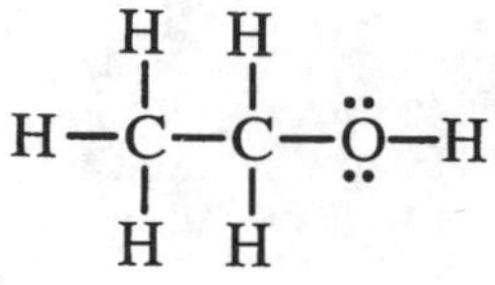

B. $C_2H_4O_2$

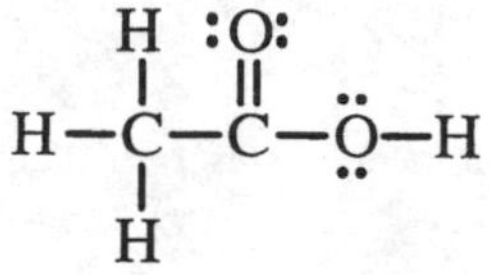

C. $C_4H_{11}N$

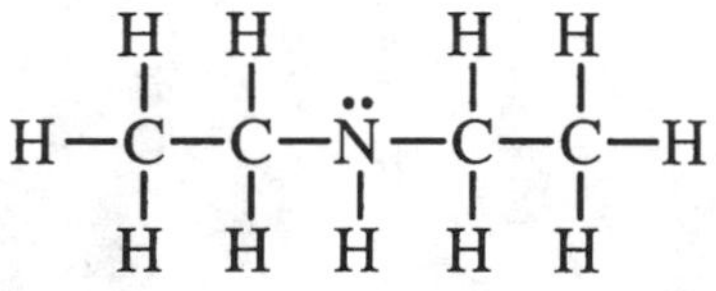

D. $C_5H_{12}O$

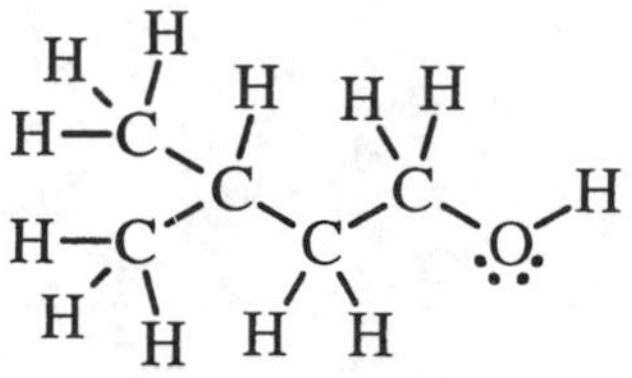

E. $C_5H_{13}N$

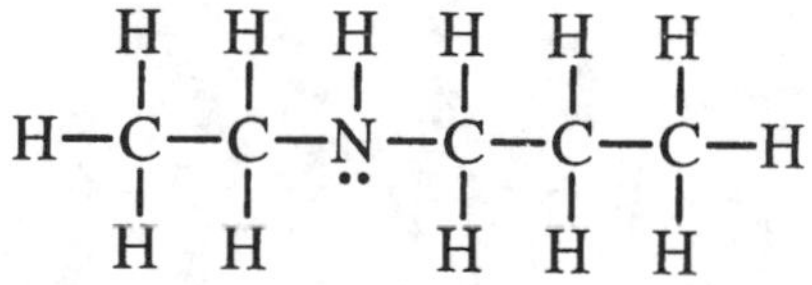

F. C_6H_{10}

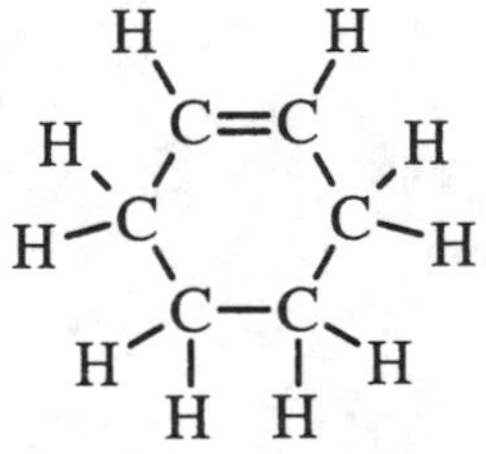

G. $C_7H_6O_3$

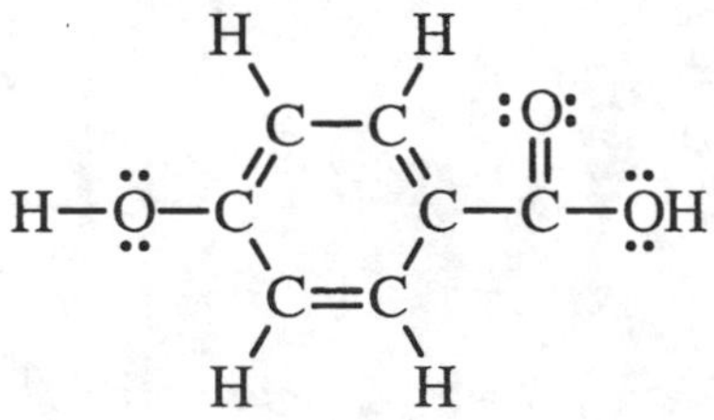

H. C_8H_{18}

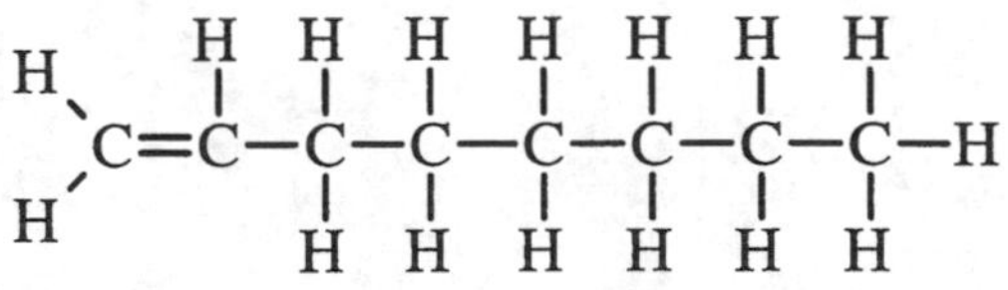

AMINES AND CARBONYLS—Exp. N-2

Name______________________________ Lab Section __________________________

Lab Partner__________________________

Problem Statement: How do organic amines react with compounds having a carbonyl group?

I. The Reaction of Aniline with Acetic Anhydride

A. About the Reactants

The Lewis Structures of the two reactants for this reaction, aniline and acetic anhydride, are given below. Aniline is called an amine because it has a nitrogen atom in it. Acetic anhydride has carbon double bonded to oxygen within it; these two atoms bonded that way are called a carbonyl group. You will react these two together in aqueous solution and identify the product by its physical properties. You will be using HCl acid to help dissolve the aniline and later adding the weak base acetate to free up the aniline and allow it to react.

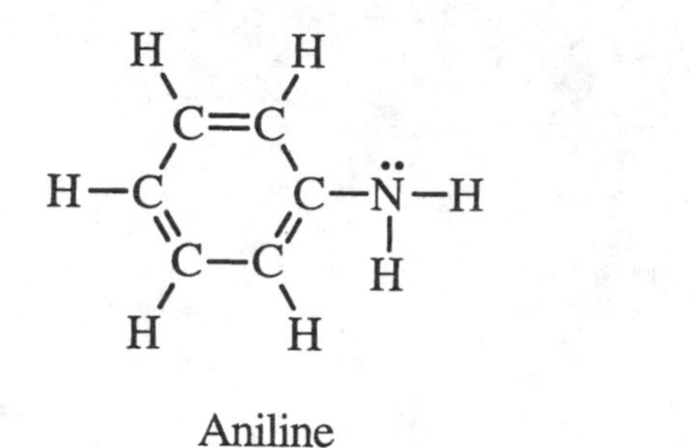

Aniline

```
   H :O:     :O: H
   |  ||  ..  ||  |
H—C—C—O—C—C—H
   |      ..      |
   H              H
```

Acetic Anhydride

B. Running the Reaction

Using volumetric cylinders, put 25.0 mL of 0.4 M HCl (aq) into a 100 mL beaker. Put 7 to 8 mL of 1.9 M $NaC_2H_3O_2$ (aq) into a separate small beaker.

Add 0.5 mL of aniline to the 25.0 mL HCl solution and stir until it is all dissolved. While one partner stirs, the other is to add 0.7 mL of acetic anhydride, then immediately add the 7 mL of 1.9 M $NaC_2H_3O_2$ (aq). Continue stirring until a change occurs. Record your observations.

C. Melting Point (Optional—see instructor for details)

Filter out the product crystals, using an aspirator suction filtration apparatus. Wash the crystals with two 1 mL portions of distilled water. Put 20 mL of distilled water into a 50 or 100 mL beaker and heat it to just boiling. Add the product crystals to the hot water and stir until dissolved. Put into an ice water bath and let sit until the compound recrystalizes. Refilter and air dry the crystals. Determine the melting point of the crystals as directed by the instructor.

II. Interpretation

A. Identify the Crystals

Consult the list of organic compounds and their properties in Table C-7 in Appendix C. What is the most likely identity of the product crystals? Give the Lewis Structure for this compound and briefly explain why you chose it.

B. A Reaction Mechanism

Using Lewis Structures and arrows, draw a picture of how aniline probably reacted with acetic anhydride. What other product most likely formed?

Using words, explain your picture above (i.e., explain how organic amines react with compounds having a carbonyl group). You might also speculate on where the electrons in the new bond came from.

III. The Reaction Between 1,6-Hexanediamine and Sebacoyl Chloride

A. About the Reactants

The Lewis Structures of the two reactants for this new reaction, 1,6-hexanediamine and sebacoyl chloride, are given below. These have been dissolved in water and in hexane, respectively, for you. These two solvents are immiscible with each other so that the reaction is forced to occur at their interface.

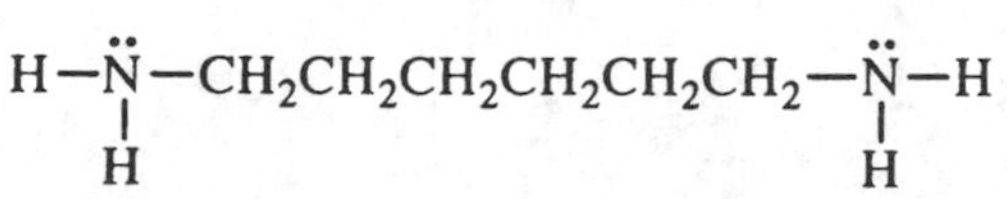

$$H-\ddot{N}(H)-CH_2CH_2CH_2CH_2CH_2CH_2-\ddot{N}(H)-H$$

1,6-Hexanediamine

$$Cl-C(=\ddot{O}:)-CH_2CH_2CH_2CH_2CH_2CH_2CH_2CH_2-C(=\ddot{O}:)-Cl$$

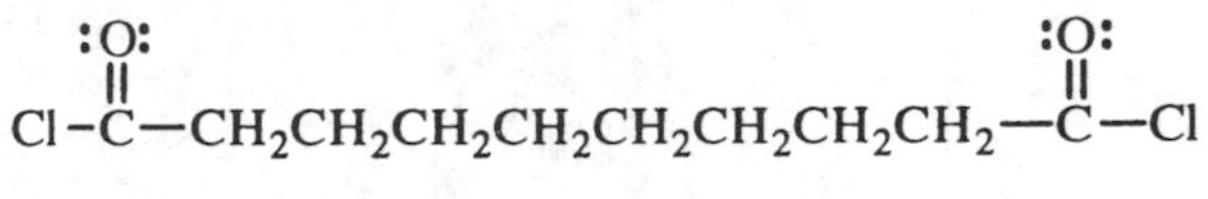

Sebacoyl Chloride

B. Running the Reaction

Put 10 mL of 0.5 M 1,6-hexanediamine into a 50 mL beaker. Get one pellet of NaOH (about 0.2 grams) and dissolve it into the 1,6-hexanediamine solution. Into a separate beaker, measure out 10 mL of 0.2 M sebacoyl chloride in hexane. Pour this gently down the side of the other solution. Do not mix. Let stand for a few minutes and record your observations.

Using tongs or forceps, grab the opaque material at the interface of the two solutions and slowly pull it up out of the solution. Continue to pull, winding the material around a stirring rod or spatula. When you have a foot or two of it, break the string and wash the wound material with water, catching the wash in a waste beaker. Investigate the texture of the material using a gloved hand. Record your observations.

Mix the remaining solution. Wash the resulting material with water, catching the wash in a waste beaker. Investigate the texture of the material using a gloved hand. Record your observations.

V. Interpretation

A. Speculate on what the chemical formula is for this new material. Use the mechanism (the reaction picture) you developed and the description you gave in part II.B to speculate on how the amine parts of 1,6-hexanediamine may have reacted with the carbonyl parts of sebacoyl chloride (i.e., give a possible Lewis Structure for the product of the reaction and briefly explain why you chose that structure).

B. Offer an explanation for the difference in physical properties of the two materials made in this experiment (i.e., why is one discrete small crystals while the other is a continuous string?).

Section Two

OPEN INQUIRY EXPERIMENTS

A. INTRODUCTION

The experiments in this section of the laboratory manual are called Open Inquiry experiments. With this section you will be designing and carrying out your own experiments much as a research scientist must do. One way to look at these Open Inquiry experiments is that they are designed to give you the time and equipment to "do your own thing," to do experiments that make sense to you and that will help you personally to learn some chemistry. The laboratory manual contains only lists of what are called systems for each general area (i.e., mass relationships, heats, etc.). These systems are only suggestions to help you get started.

There are two classroom approaches to doing Open Inquiry using these systems.

I. Individual Approach

Here, you are free to work with your partner on any system you wish. You may work on one of the systems given, or invent your own. The only limitations are: (1) that the investigation should be connected to the content of the Guided Inquiry (Section One) experiment previously done, (2) that the equipment and chemicals necessary for the investigation are available or easily constructed (see your lab instructor), and (3) that your safety and the safety of others is safeguarded (make sure your laboratory instructor knows what you are doing).

There are no detailed instructions on how to approach these systems. This will be your job. There are many different ways to approach each system. The choice is yours. Following is a list of ground rules and suggestions that you should observe:

a. You must complete the assigned Guided Inquiry experiment (Section One) laboratory report before beginning your Open-Ended experiments.

b. If you have no ideas of your own on what problem to investigate, read through the systems for that general area. If you are still undecided as to what to do, select a couple of the systems and do some short, qualitative investigations of them. From your observations you should find some aspect of a system to investigate in more detail.

c. Before beginning your detailed investigation it would be helpful to make a rough outline of the procedures you will use, i.e., what chemicals and equipment you will need, what you will measure, what you will vary, what you will hold constant. ***Check with your lab instructor to be sure your experiments are reasonably safe.***

d. You should spend most of the lab time available in actual experimental work. If you finish an investigation to your satisfaction and there is still lab time remaining, start investigating another system.

II. Whole Class Approach

Here the whole class will work on a single system, gathering data on its different facets and sharing that data to get a more complete picture of the chemistry. In this approach your responsibility will be to contribute thoughtfully to the class brainstorming and discussions, to help design effective experiments, and to gather reproducible data bearing on the question. The general flow of the investigation will be as follows:

a. All pairs do qualitative experiments on the chosen system to get some experience with it.

b. The full class brainstorms hypotheses, chooses some to investigate, and then brainstorms experiments that will yield worthwhile data.

c. The class divides up the possible experiments and each student pair does those assigned, reporting the data with 30 minutes left in the period.

d. The class discusses the meaning of the individual data sets and the conclusions that can be drawn from the collective set. Individuals then write up their own reports on whether the hypotheses were substantiated or disproved.

III. Laboratory Report

The laboratory report on your Open Inquiry experiments should be written so that any person could follow it. Try to be both concise and clear. The report should be divided into the four parts listed below.

a. ***Problem Statement.*** This should be just a few sentences stating the problem or describing the system you investigated. If you had any predictions as to what would happen, include them here.

b. ***Procedures.*** This should be a brief outline of the experimental procedures you used, including the equipment and chemicals employed and what variables in the experiment were measured. A detailed step-by-step account of the experiments is not required.

c. ***Data Presentation and Analysis.*** Data should be presented in organized tabular form and labeled so it is easy for a reader to follow. Algebraic equations, mathematical relationships, and graphical representation of the data should be used when possible and appropriate. All of these should be developed and presented in a way so that their connection and relevance to the data are clear.

d. ***Interpretations and Conclusions.*** This part should include your conclusions, explanations and any generalizations you feel you can make about the chemistry. The logic you used in arriving at these ideas should be clearly stated. Try to also include logical alternative explanations and hypothesis.

Your grade will be derived from how you investigated your system, how thoroughly you investigated the system in the time allowed, how the lab report and data are presented, and the reasonableness of your data analysis and explanations. It is especially important that all of your conclusions are logically consistent with the data you collected. Although you are encouraged to work with your partner, and to share ideas and information with others, you will be expected to write up your final report independently.

Michael R. Abraham
Michael J. Pavelich

B. MEASUREMENT SYSTEMS

System 1

Investigate the uncertainty of measurement instruments. Possibilities include:

cylinders
meter sticks
thermometers

System 2

Investigate the uncertainty involved in determining the volume of a regularly shaped solid.

System 3

Investigate the uncertainty of measuring an object with a single instrument as compared with using several different examples of the instrument.

System 4

Investigate the physical relationships of metal cylinders made of metals other than brass or aluminum. Compare their physical relationships with brass and aluminum.

System 5

Investigate the graphical relationship between the temperature of a cooling liquid and the time passed.

System 6

Investigate, by graphing, the relationship between two variables looked up in a handbook. Possibilities include:

density and melting point of materials (metals)
density and heat capacity of materials

System 7

Investigate the volume of water and its height in various containers (i.e., funnel or flask).

System 8

Investigate the uncertainties associated with the weight and density of pennies. What are the sources of this uncertainty?

System 9

Investigate the measurement relationships associated with any other system or any modification of any of the above systems. For safety reasons, discuss your system with the instructor before proceeding.

C. MASS RELATIONSHIPS SYSTEMS

System 1

Investigate the mass relationships associated with heating various compounds. Possibilities include:

magnesium—Mg
boric acid—H_3BO_3
sodium bicarbonate—$NaHCO_3$
copper (II) carbonate—$CuCO_3$

System 2

Investigate the mass relationship associated with heating hydrated chemical substances. One or more of the following are possibilities:

hydrated ferrous sulfate
hydrated barium chloride
hydrated cobaltous chloride
hydrated copper (II) chloride

System 3

Investigate the mass relationships associated with adding 4 M hydrochloric acid (HCl) dropwise to sodium carbonate (Na_2CO_3) or to sodium bicarbonate ($NaHCO_3$). **CAUTION!!!** See System 6.

System 4

Investigate the mass relationships associated with heating a mixture of sulfur and a metal. (**CAUTION!!!** Obnoxious odor may form. Use a hood or an aspirator-funnel arrangement). Possible metals are:

copper
iron
lead

System 5

Investigate the mass relationships associated with the formation of precipitates when a solution of one ionic compound is mixed with a solution of another ionic compound. Possibilities include:

lead (II) nitrate and potassium iodide
silver nitrate and potassium chloride
calcium chloride and sodium carbonate
copper (II) sulfate and sodium phosphate
iron (III) nitrate and sodium phosphate
nickel (II) sulfate and potassium hydroxide

System 6

Investigate the mass relationships associated with the addition of metals to diluted solutions of hydrochloric acid (HCl) or sulfuric acid (H_2SO_4). (Hints: Dilute concentrated acids by adding acid to water. The strength of acid solutions can be determined by titrating with a standardized NaOH solution.) **CAUTION!!!** HCl, H_2SO_4 and NaOH are corrosive to the skin. Wash spills with large amounts of water. Possible metals include:

iron
zinc
magnesium
aluminum

System 7

Investigate the mass relationships associated with any other system or any modification of any of the above systems. For safety reasons, discuss your system with the instructor before proceeding.

D. HEAT LAWS SYSTEMS

System 1

Investigate the heat lost or gained when a specific chemical or group of chemicals are added to water. The following are possibilities:

sodium nitrate
sodium chloride
hydrated calcium chloride
urea
ammonium nitrate
sodium hydroxide (**CAUTION!!!** This material is corrosive and should not be allowed to come in contact with your skin.)

System 2

Compare the heat lost or gained by hydrated versus anhydrous compounds when they are dissolved in water. It may be necessary to make your own anhydrous salts from the available hydrate. Possibilities include:

$MgSO_4(H_2O)_7$ vs. $MgSO_4$
$CaCl_2(H_2O)_2$ vs. $CaCl_2$
$CuSO_4(H_2O)_5$ vs. $CuSO_4$
$BaCl_2(H_2O)_2$ vs. $BaCl_2$

System 3

Investigate the heat gained or lost when various bases are mixed with various acid solutions. For safety reasons use acid concentrations of 2.0 M or less. The bases can be used in solid form or in solution. Possible chemicals include:

Bases	**Acids**
NaOH	HCl
KOH	H_2SO_4
$Ca(OH)_2$	HNO_3

System 4

Investigate the effect of changing the calorimeter conditions on the heat of a solution reaction. See Systems 1–3 for possible reactions. Calorimeter variables include:

glass vs. metal vs. polystyrene container
open vs. closed calorimeter
amount of water
initial temperature of water
ice present in the water
thickness of the polystyrene walls

System 5

Investigate the heat lost or gained when a "hot" or "cold" piece of metal is added to water. The following are possibilities:

aluminum
zinc
brass

System 6

Investigate the temperature changes occurring during a phase change. The following are possibilities:

melting of ice
melting of wax

System 7

Investigate the temperature of the air and other gases above water as the water is heated to boiling.

System 8

Investigate the melting point of mixtures of substances. Try lauric acid doped with small amounts of benzoic acid.

System 9

Investigate the heats of chemical reactions. The following are possibilities:

citric acid ($H_3C_6H_5O_7$) and baking soda ($NaHCO_3$)
HCl(aq) and Mg(s)

System 10

Investigate the heat lost or gained with any other system or any modification of any of the above systems. For safety reasons, discuss your system with the lab instructor before proceeding.

E. GAS SYSTEMS

System 1

Investigate the behavior of gases other than air. Possibilities include oxygen (can be generated by heating potassium chlorate with a pinch of manganese dioxide as a catalyst), or carbon dioxide (can be generated by adding dilute acid to calcium carbonate). **CAUTION!!!** When diluting acids extreme care must be taken not to allow the acid to come into contact with your skin or clothes. Also add acid to water, *not vice versa*, when diluting.

System 2

Investigate and compare the vapor pressures of pure liquids with combinations of these liquids. See your instructor for suggested liquids.

System 3

Investigate the relationship between the amount of gas and its pressure at constant temperature and volume.

System 4

Investigate and compare the vapor pressures of liquids using a 250 mL suction flask instead of a 125 mL flask.

System 5

Investigate and compare the vapor pressure of pure liquids with solutions of those liquids containing measured amounts of dissolved solid chemicals.

System 6

Investigate the relationship between the density of air and temperature. (Hint: Derive an equation relating the change in mass of air at different temperatures at constant pressure and volume.)

System 7

Determine the molecular weight of gases from the mass of a measured volume. CO_2 and CH_4 are examples.

System 8

Investigate any other gas system or investigate a modification of any of the above systems. For safety reasons, discuss your system with the lab instructor before proceeding.

F. PERIODICITY SYSTEMS

System 1

Look up the specific properties of a series of compounds or elements. By graphing, look for patterns in the properties versus position of an element in the Periodic Table and propose an explanation for the patterns found. Ideas on what properties and species might be used are given below. Information sources are *Handbook of Chemistry and Physics*, computer software that documents properties of the elements, and/or your own textbook.

a. solubility of salts like: NaF, NaCl, NaBr, NaI, and/or LiBr, NaBr, KBr, RbBr
b. density of salts like: NaF, NaCl, NaBr, NaI, and/or NaCl, $MgCl_2$, $AlCl_3$, $SiCl_4$
c. melting points of compounds like: NaF, NaCl, NaBr, NaI, and NaCl, $MgCl_2$, $AlCl_3$, $SiCl_4$
d. acidity of compounds like: HF, HCl, HBr, HI, and/or H_2O, H_2S, H_2Se, H_2Te
e. ionization energies of the Main Group (A Group) elements
f. atomic radii of the Main Group (A Group) elements
g. ionic radii for the group 1A, 2A, 3A cations, and 6A, 7A anions

System 2

Investigate the ionic nature of a series of substances. (Hint: The ionic nature of a substance can be related to its electrical conductivity when dissolved in water. This property can be measured qualitatively with a "light bulb" apparatus, and quantitatively with a conductance meter. See your laboratory instructor for instructions on their use.)

a. To be measured qualitatively:

H_2O (distilled)	magnesium chloride, $MgCl_2$
sodium chloride, NaCl	sugar, CH_2O
sodium bromide, NaBr	potassium chloride, KCl
cyclohexane, CH_2	potassium bromide, KBr
potassium chloride, KCl	

b. To be measured quantitatively (Hint: Measure the resistance of very weak solutions; no more than 0.0025 moles of solid in 250 mL of solution.):
   1. Nitrate salts of Li^+, Na^+, K^+, Mg^{2+}, Ca^{2+}, Sr^{2+}, Ba^{2+}, Al^{3+}
   2. Sodium salts of F^-, Cl^-, Br^-, I^-

System 3

Investigate the reactions between solutions of a series of cations with solutions of an anion. A possible series of cations includes:

Mg^{2+}, Ca^{2+}, Sr^{2+}, Ba^{2+} as a water soluble nitrate, chloride, etc.

One or more of the following are possible anions.

CO_3^{2-} [$(NH_4)_2CO_3$]
SO_4^{2-} (dilute H_2SO_4)
OH^- (as NaOH and as NH_4OH)

System 4

Investigate the chemical properties of any other series of elements or ions or investigate a modification of any of the above systems. For safety reasons, discuss your system with the lab instructor before proceeding.

G. ACID AND BASE SYSTEMS

System 1

Investigate the pH properties of acids and their "conjugates." Below are possible acids to investigate:

Acid	Conjugate
$HC_2H_3O_2$	$NaC_2H_3O_2$
HCl	$NaCl$
$H_2C_2O_4$	$Na_2C_2O_4$
H_2SO_4	Na_2SO_4
H_3PO_4	Na_3PO_4

System 2

Investigate the neutralization of an oxalic acid solution

$$\mathrm{H-O-\underset{\underset{O}{\|}}{C}-\underset{\underset{O}{\|}}{C}-O-H\,(s)} \xrightarrow[\mathrm{H_2O}]{\text{in}} \mathrm{2H^{+}(aq) + {}^{-}O-\underset{\underset{O}{\|}}{C}-\underset{\underset{O}{\|}}{C}-O^{-}(aq)}$$

when it reacts with a standardized NaOH solution.

System 3

Investigate the pH properties of basic compounds when they interact with the strong acid HCl in solution. Possible bases are:

Na_2HPO_4
Na_3PO_4
$NaOH$
Na_2CO_3
NH_4OH

System 4

Investigate the pH at which different indicators change color. (Hint: Titrate 25 mL of a 0.1 M acetic acid solution with a standardized base and/or titrate 25 mL of a 0.1 M Na_2CO_3 with 0.1 M HCl.) Below are possible indicators to use:

phenolphthalein	methyl orange
bromocresol green	indigo carmine
alizarin yellow	methyl red

System 5

Investigate the acid-base properties of some common household products.

a. Investigate foods and household chemicals like orange juice, 7-Up (fresh and "flat"), vinegar, ammonia, etc.

b. Investigate the power of various commercial antacids.

c. Investigate the power of various brands of aspirin. (Hint: The active drug in aspirin is acetylsalicylic acid.)

System 6

Investigate the pH and conductivity properties of a set of acids or bases. (Hint: For best results solutions should not exceed 0.01 M). Possible acids and bases are:

Acids	Bases
HCl	$NaOH$
H_2SO_4	NH_3
H_3PO_4	Na_3PO_4
$HC_2H_3O_2$	$NaC_2H_3O_2$
$H_2C_2O_4$	$Ba(OH)_2$
KHP	

System 7

Investigate the acid-base properties of any other system or investigate a modification of any of the above systems. For safety reasons, discuss your system with the lab instructor before proceeding.

I. EQUILIBRIUM SYSTEMS

System 1

Investigate the solubility of lead salts such as

lead (II) chloride—$PbCl_2$
lead (II) bromide—$PbBr_2$
lead (II) oxalate—PbC_2O_4

These can be made by reacting $Pb(NO_3)_2$ solutions with solutions of NaCl, NaBr or $Na_2C_2O_4$.

System 2

Investigate how I_2 distributes itself between water and C_6H_{12} (cyclohexane) when the two immiscible (insoluble) liquids and I_2 are mixed together.

NOTE: The concentration of I_2 in each liquid can be determined by titrating with a sodium thiosulfate ($Na_2S_2O_3$) solution of known concentration. The titration (analysis) reaction is:

$$I_2 + 2\,S_2O_3^{2-}(aq) \rightarrow 2I^-(aq) + S_4O_6^{2-}(aq)$$

System 3

Investigate the "completeness" of precipitation reactions. The following are possibilities:

$Co(NO_3)_2$ and Na_2CO_3
$NiSO_4$ and Na_2CO_3
$Co(NO_3)_2$ and $Na_2C_2O_4$

System 4

Investigate the pH properties of acids. Possibilities include:

potassium acid phthalate—$HKC_8H_4O_4$
acetic acid—$HC_2H_3O_2$
oxalic acid—$H_2C_2O_4 \cdot 2H_2O$
hydrochloric acid—HCl

System 5

Investigate the pH properties of bases. Possibilities include:

sodium acetate—$NaC_2H_3O_2$
sodium carbonate—Na_2CO_3
potassium hydroxide—KOH
sodium phosphate—Na_3PO_4
sodium monohydrogen phosphate—Na_2HPO_4
sodium oxalate—$Na_2C_2O_4$
sodium chloride—NaCl

System 6

Investigate the effect on pH of mixing an acid with a base. See systems 4 and 5 for suggestions.

System 7

Investigate the pH and conductivity properties of a set of acids or bases. (Hint: For best results solutions should not exceed 0.01 M). Possible acids and bases are:

Acids	**Bases**
HCl	$NaOH$
H_2SO_4	NH_3
H_3PO_4	Na_3PO_4
$HC_2H_3O_2$	$NaC_2H_3O_2$
$H_2C_2O_4$	$Ca(OH)_2$
KHP	

System 8

Investigate the concentration properties of any other system or modification of any of the above systems. For safety reasons, discuss your ideas with the lab instructor before proceeding.

J. ELECTROCHEMICAL SYSTEMS

System 1

Investigate the effect of decreasing the concentration of one of the ½ cells of the Cu/Cu^{2+} | | Cu^{2+}/Cu cell.

System 2

Investigate the properties of other cell combinations. (Note that some ½ cells contain more than one dissolved species.) Possible half cells include:

Mg/Mg^{2+} | |
Fe/Fe^{2+} | |
Pt/Mn^{2+}, MnO_4^-, H^+ | |
Pt/Cr^{3+}, $Cr_2O_7^{2-}$, H^+ | |
Ni/Ni^{2+} | |
Sn/Sn^{2+} | |

System 3

Investigate the effect of pH on the voltage of various cells using Pt/CrO_4^{2-}, $Cr_2O_7^{2-}$ | | as a ½ cell, with other ½ cells.

System 4

Investigate the effect of amount of added $NH_3(aq)$ on the voltage of various cells using Ag/Ag^+ or Cu/Cu^{2+} as a ½ cell with other ½ cells.

System 5

Investigate the use of other metals besides Cu in the electrolysis described in Exp. J-2, part VI.

System 6

Investigate the electrolysis of ionic salt solutions using the following apparatus. Possible salts to investigate include:

$CuCl_2$
KI

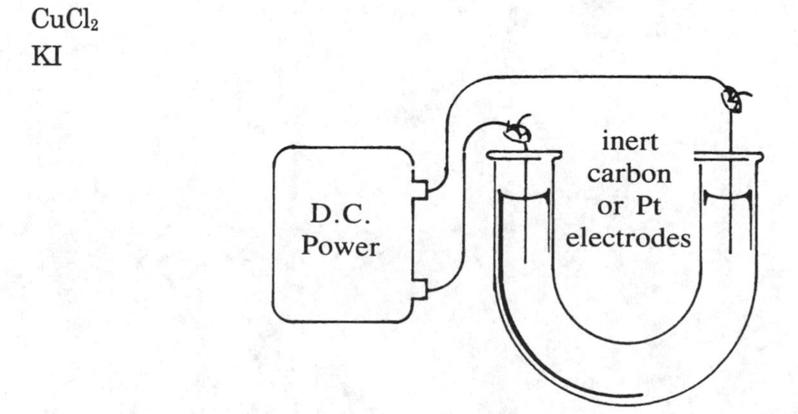

System 7

Investigate the electrolysis of a metal in a metal ion solution using the following apparatus. Possibilities include Cu and Pb.

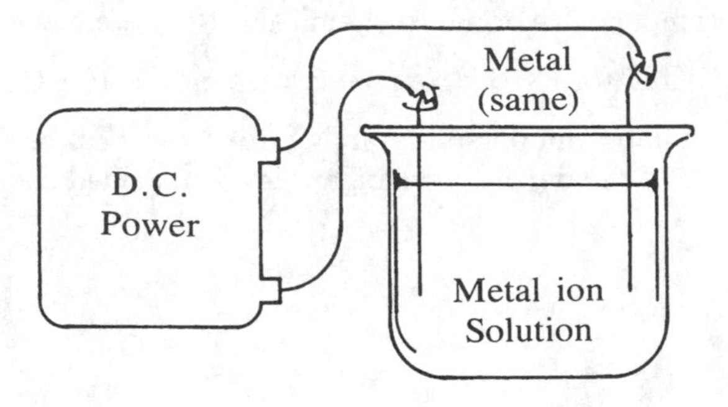

System 8

Investigate the electrolysis of a voltaic cell. Use the following apparatus.

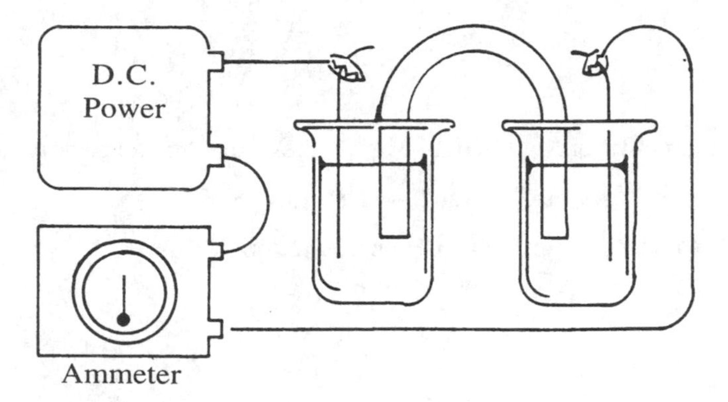

System 9

Investigate combinations of voltaic cells.

System 10

Investigate the electrochemical properties of any other system or any modification of any of the above systems. For safety reasons, discuss your system with the lab instructor before proceeding.

K. KINETICS SYSTEMS

System 1

(computer use?)

Investigate how concentrations, temperature or added chemicals affect the rate of the reaction:

$$2\ MnO_4^-(aq) + 5\ C_2O_4H_2(aq) + 6\ H^+(aq) \rightarrow 2\ Mn^{+2}(aq) + 10\ CO_2(aq) + 8\ H_2O$$

See Exp. K-1 for solution concentrations and possible starting mixtures. Can be followed visually or spectrophotometrically as the disappearance of the purple permanganate color. Added chemicals can be solids like Fe, Cu, MnO_2, SiO_2 (sand).

System 2

(computer use?)

Determine the rate law for the reaction:

$$(CH_3)_2CO(aq) + I_2\ (aq) \rightarrow ICH_2COCH_3 + HI(aq)$$

The reaction can be followed spectrophotometrically at 353 nm. Use stock solutions of 4 M acetone, 1 M HCl and a solution that is 0.001 M I_2 in 0.1 M KI. Start with a 1:1:1 mixture of the three.

System 3

(computer use?)

Investigate the effect of $[H^+]$, amount of Mg, form of Mg, temperature on the reaction:

$$Mg(s) + 2\ H^+(aq) \rightarrow Mg^{+2}\ (aq) + H_2(g)$$

Use [HCl] in the range of 0.1 to 0.02 M and follow pH as a function of time.

System 4

Investigate the kinetic behavior of the "clock reaction." Two possibilities are given below.

$IO_3^- + 3\ HSO_3^- \rightarrow 3\ HSO_4^- + I^-$	(use dilutions of stock solutions "A" and "B")
$HCHO + SO_3^{2-}\ H_2O \rightarrow CH_2(OH)SO_3^- + OH^-$	(use dilutions of stock solutions "C" and "D" with phenolphthalein indicator)

System 5

Investigate the effect of $[H^+]$, amount of Fe, form of Fe, temperature on the reaction:

$$Fe\ (s) + 2\ H^+(aq) \rightarrow Fe^{2+}\ (ag) + H_2(g)$$

(You can use the disappearing color of permanganate ion since the following reaction is fast:)

$$5\ Fe^{2+} + MnO_4^- + 8\ H^+ \rightarrow 5\ Fe^{3+} + Mn^{2+} + 4\ H_2O$$

System 6

Investigate the kinetic behavior of any other system or any modification of any of the above systems. For safety reasons, discuss your system with the lab instructor before doing the experiments.

APPENDICES

APPENDIX A

SAFETY IN THE LABORATORY

Any chemical laboratory by its very nature has several potential hazards which can cause injuries such as chemical burns, fire burns, poisoning and cuts. However, serious injuries rarely occur in student labs when safety rules such as the following are thoughtfully and strictly observed.

Safety Precautions

1. **Wear Safety Goggles at All Times**. Injury to the eyes can easily cause a lifetime of blindness. Safety goggles such as those pictured offer excellent protection against such injury. These must be worn over the eyes (not on your forehead!) at all times that anyone in the lab has an experiment going.

2. **Wear Proper Clothing**. To protect the feet and legs from chemical spills, wear shoes (not sandals) and full-length, loose-fitting pants in the lab. Also it is advised that old clothes be worn as minor chemical spotting will occur from time to time. Tie back long hair so that it does not fall into flames or chemicals.

3. **Do Not Touch, Taste or Smell Lab Chemicals**. Use the proper equipment (spatulas, pipets, beakers—see Appendix B) to handle chemicals. Tasting of lab chemicals is strictly forbidden, as is eating or drinking normal foods in the lab. Any of these can easily lead to poisoning. If you must smell a chemical, use an indirect wafting technique.

4. **Do Not Use Cracked or Chipped Glassware**. Replace such glassware or fire polish chips at the direction of your instructor.

5. **Dispose of Chemicals as Directed**. Do not put lab chemicals into waste baskets or directly down a sink drain. Waste chemicals should be placed in crocks or disposal containers as directed by the instructor. (Occasionally, the instructor may allow you to wash soluble, harmless chemicals down the drain.)

6. **Know the Location and Use of Safety Equipment**. Your lab will be equipped with safety equipment such as the items pictured on the following page. Talk to your instructor about how and when they are to be used. Find at least two ways to exit the lab.

7. **Never Leave a Burner Flame or Reaction Unattended**.

8. **Follow Safety Instructions for Your Specific Experiment to the Letter**. Read Appendix B and other appendices as needed. Discuss the procedures with the instructor if they are unclear to you.

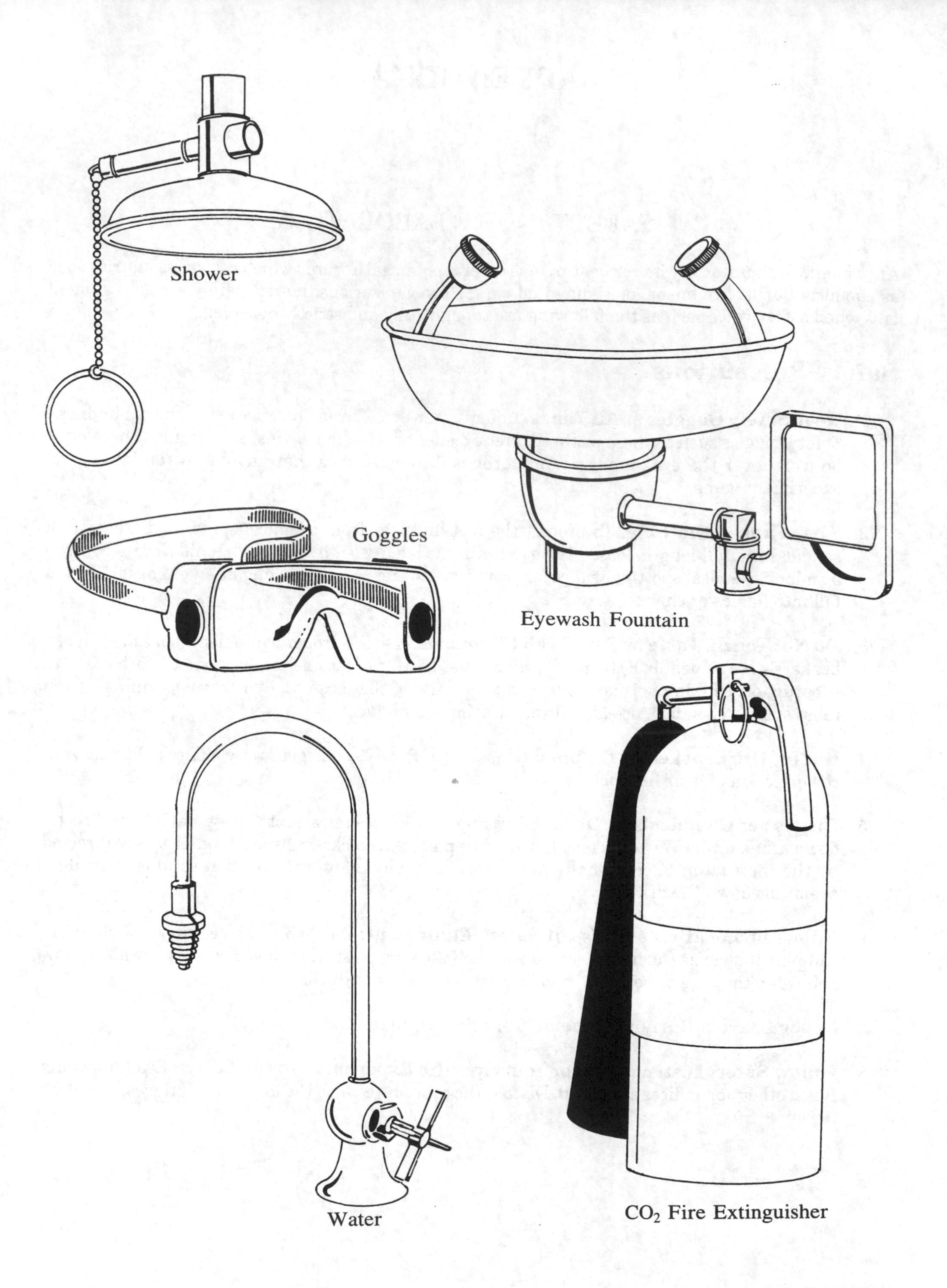
Shower
Goggles
Eyewash Fountain
Water
CO2 Fire Extinguisher

WHAT TO DO IN CASE OF ACCIDENTS

Minor Accidents

1. **Skin Contact with Chemicals.** This is the most common form of accident, usually affecting the hands and forearms. The key word for treatment is **water,** large volumes of it. **Immediately** wash the affected area with large volumes of water. Do this even if no sensation of burning is felt, since many chemicals are slow acting. Use a faucet for the arms and hands.

   For the face, use an eye-wash fountain. Keep your goggles in place for the beginning of the wash so as not to wash the chemical into your eyes. For the torso or legs go to the restroom, disrobe as needed, and wash the affected area with large amounts of water. Also rinse the affected clothing before using it again. **If an immediate, heavy burning sensation is felt on the torso or legs, use the safety shower.** Report the accident to the instructor.

2. **Chemical Spills.** Any spills, no matter how small, on the bench or floor must be taken care of immediately to prevent others from unknowingly getting injured by them. Check with the instructor for clean-up procedures.

   Usually a spill can be handled by wiping it up with a wet cloth or sponge. The spot should then be washed twice more with fresh water and wiped dry.

3. **Heat Burns.** These come from the momentary touching of hot pieces of metal or glass. Hold the burned area under a stream of cold water or pack it with ice. Report the injury to the instructor.

4. **Cuts**. For small cuts, wash them with water and secure a dressing from the instructor.

5. **Small Fires.** These can usually be extinguished by smothering with a double thickness of toweling.

6. **Think, then Act.** Most injuries occur from reacting too fast to a situation; i.e., an overflowing reaction, a falling beaker. (Trying to catch a beaker of corrosive chemical is tricky, to say the least.) The best thing to do in such a situation is to shout a warning, back away, and let the accident happen. Then think a few seconds about what you can safely do to rectify the situation—and do it. Normally, this will simply be to get the instructor.

Serious Accidents

You be the judge of what is serious. Remember it is better to be slightly embarrassed than to let a situation worsen. Your immediate reaction should be to:

1. Back away from the danger area.
2. Shout for help.
3. Try not to panic; stay as calm as possible. The instructor will know what to do.

If you are not involved in the accident, you may be called on to lend your instructor some assistance. Primarily this assistance will take the form of getting a person under the safety shower to extinguish flames or to wash off a serious spill.

APPENDIX B

COMMON PROCEDURES AND EQUIPMENT

I. Balances

The weighing (or massing) of chemicals is one of the most used and potentially most precise measurements employed by chemists. The measurement is done by ***counterbalancing*** the substance of unknown mass against objects of known mass. This balancing act is accomplished by means of a delicate pivot point called a knife edge. Thus the measuring instruments used are called ***balances***.

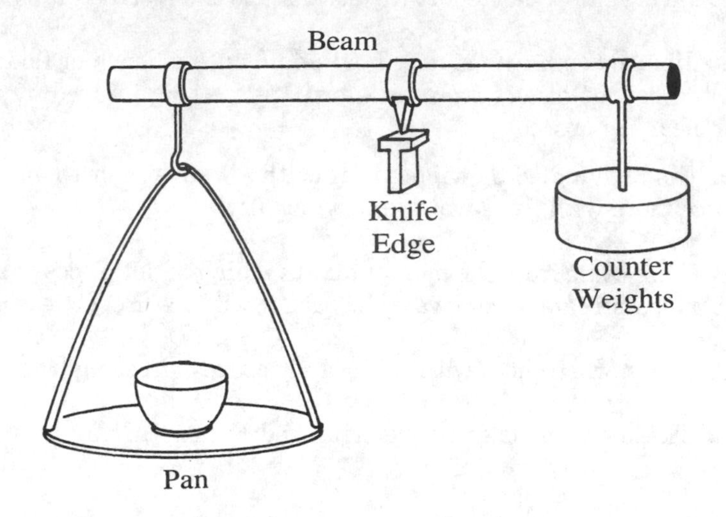

Schematic of a Balance

Three types of balances are pictured and discussed here: the triple beam balance, the top loading electronic balance and the analytical electronic balance. Your instructor will give you specific instructions for the type of balance you will use; however, it is strongly suggested that you study the comments below before doing any balance work.

A. CARE OF THE BALANCE

Because balances are very delicate and fairly expensive instruments, the following rules must be followed.

1. Do not attempt to repair a balance that is malfunctioning. Inform your instructor about the problem.
2. Never place chemicals directly on the balance pans, as this will lead to corrosion and eventual malfunctioning. Beakers, paper "boats," etc., should be used to hold the chemical being weighed.
3. Never remove chemicals from a container while it is on the balance pan. This might put too much pressure on the pan. Always remove the container completely from the balance before removing chemicals.

4. Add chemicals to a container on a balance pan carefully, to prevent spills. If a spill does occur, use a balance brush to gently sweep it away. (Be sure the electronic balance is shut off before you start cleaning up.) If the spill involves more than a few grains of material or involves a highly corrosive chemical, see your instructor for clean-up instructions.
5. Do not move a balance. This will upset its leveling and can cause the machinery to jam.

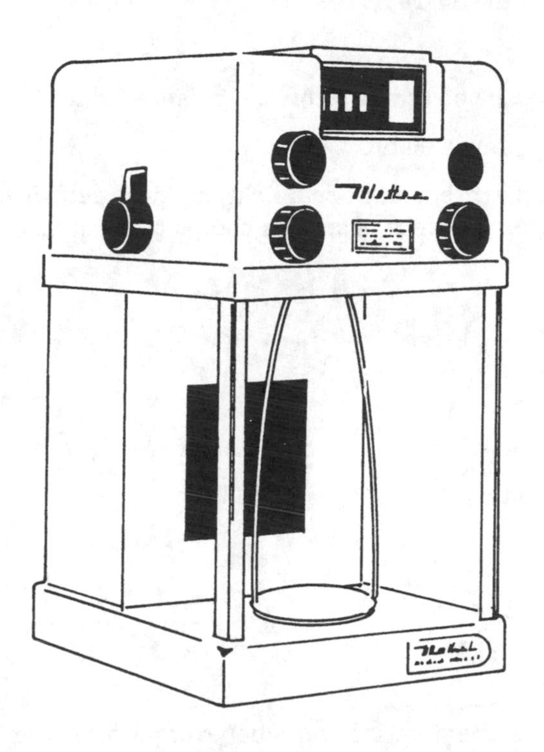

Top-loading Electronic Balance

Analytical Electronic Balance

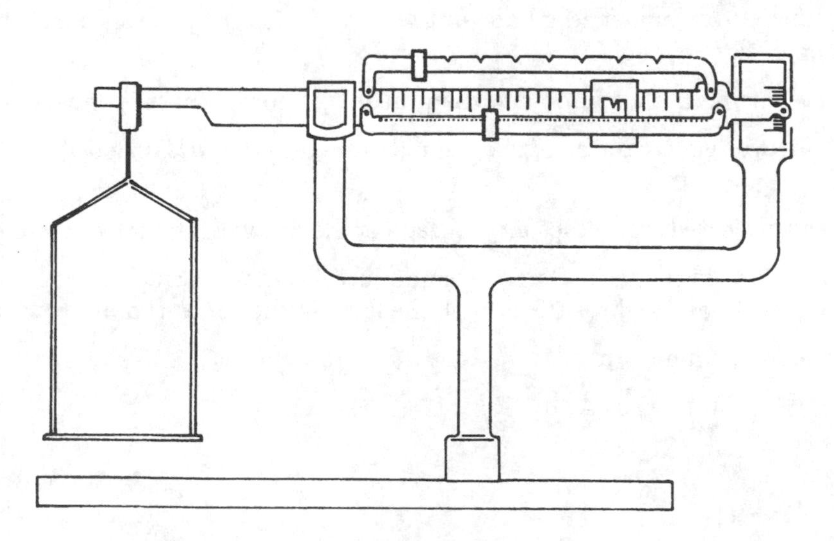

Triple Beam Balance

B. USE OF THE ANALYTICAL ELECTRONIC BALANCE

(The analytical balance discussed below is the older, partially mechanical version still being used in many laboratories.)

This machine ***looks*** totally electronic and fool-proof because all that shows is a balance pan, some knobs and a weight read-out window. However, it is basically like other balances; the knife-edge, balance beam and counterweights are just enclosed in the housing. The balance is very precise (± .001 to ± .0002g), also making it very delicate and very easily damaged. Thus it must be used thoughtfully and gently.

Your instructor will explain the use of your specific balance. In general it has four types of controls:

1. Zeroing knob
2. Large weight knobs—turning these causes weights to be placed on or off the balance beam.
3. Fine weight knob—this allows for a better estimate of mg weights.
4. On-off lever—this has three positions; ***off*** or ***full arrest*** (balance beam held in place), ***partial arrest*** (balance beam has limited movement), ***full on*** (balance beam has complete freedom of movement).

ZEROING

1. Check that the instrument is ***off.*** Adjust all weight knobs to zero and close the doors.
2. Turn to ***full on*** and adjust to read 0.0000g by gently turning the zeroing knob.
3. Turn to ***off.***

WEIGHING

1. Place the container to be weighed on the pan. The instrument must be ***off*** whenever placing hings on or removing them from the pan.
2. Turn lever to ***partial arrest*** position.
3. Gently and slowly adjust 10 g knob until the projected scale moves abruptly upward. Then turn the knob back one step.
4. Repeat #3 using the 1 g knob and then the 0.1 g knob, if there is one on your instrument.
5. Turn the lever gently to the ***off*** position and then to the ***full on*** position. Be sure the doors are closed.
6. When the projected scale stops moving, adjust the fine weight knob as directed by your instructor.
7. Read and record the mass. (Some machines use a comma as a decimal point in the gram weight.) Be sure your final weight is taken with the lever in the ***full on,*** not the ***partial arrest*** position.
8. Turn the instrument ***off.***

C. USE OF THE TOP LOADING ELECTRONIC BALANCE

(The balance discussed below is the newest, totally electronic version.)

This instrument uses a pressure sensitive strain gauge to determine mass. The mass of the object affects the electronic signal passing through the gauge; this signal gets displayed as a digital readout of grams. The instrument can be as precise as many analytical balances (±0.001 g) and is easier to use. Such top loading balances are becoming the instrument of choice for routine lab work. However, they are delicate and expensive to repair. Thus they must also be used thoughtfully and carefully.

The controls will be explained by your instructor but, in general, they are as follows:

1. on-off button
2. calibration button
3. tare button

WEIGHING

1. Turn the instrument on using the on-off button.
2. Push the calibration button. The readout will rapidly run through a series of displays as the electronics are calibrated. When the display settles on 0.00 g, the calibration is complete.
3. Place the object to be weighed on the pan, wait a few seconds for the display to settle down, then read the gram mass of the object.
4. Your instructor will tell you either to turn off the balance after you complete your weighing or to leave it on. Follow these directions.

TARING

The tare button allows you to reset the display to 0.000 g even with an object on the balance pan. It is used when you ***only*** need the mass of the chemical and not the mass of the chemical and container. Use the tare feature only when you are absolutely sure you will never need to know the mass of the container. If you are unsure, ignore the tare button and record all masses.

1. Weigh the container as described above. After the display settles down, push the tare button. The display should now read 0.000 g.
2. Add the desired chemical to the container. The display now shows the weight of only the added chemical.

D. USE OF THE TRIPLE BEAM BALANCE

With this instrument, all the essential parts of the balance are visible—the knife edge, balance beam and counterweights. It can give a precision of ±0.02g if handled properly. Be careful not to move the balance arm sideways (by bumping it) as this might move the knife edge out of position.

ZEROING

1. With nothing on the pan, set all counterweights to their zero position. Be sure your fingers are clean and dry before you handle the weights. It is best to use a spatula or a pencil eraser to move the 0.1 g sliding counterweight.

2. When zeroed, the pointer will swing an equal distance on either side of the center line or it will rest at the center line. If this is not the case, adjust the screw nuts at the extreme left-hand side of the balance.

WEIGHING

Place the container to be weighed on the pan. Using the instructions given in #1 above, move the counterweights until the pointer again rests on or makes equal swings around the center line. Be sure the larger weights are properly seated in the notches on the arms. Read and record the total mass, then return the weights to their zero positions.

E. TIPS ON PRECISE WEIGHING

To obtain the most precise (reproducible) results from your weighing, the following procedures should be observed. These are particularly important when using the high-precision analytical balance.

1. Check the leveling bulb on the instrument. If it is not dead center, have your instructor make the needed adjustments.
2. Always use the same balance in weighing for a particular experiment. This will allow any small constant error in the balance to be canceled out in your final calculations.
3. Recheck the zero adjustment of the balance occasionally.
4. Be sure the balance and the area around it are kept spotlessly clean.
5. Never use your fingers to directly handle a container you are weighing. Fingerprints weigh a few mg each and slowly evaporate. Handle your containers with tongs or with a paper band as shown.

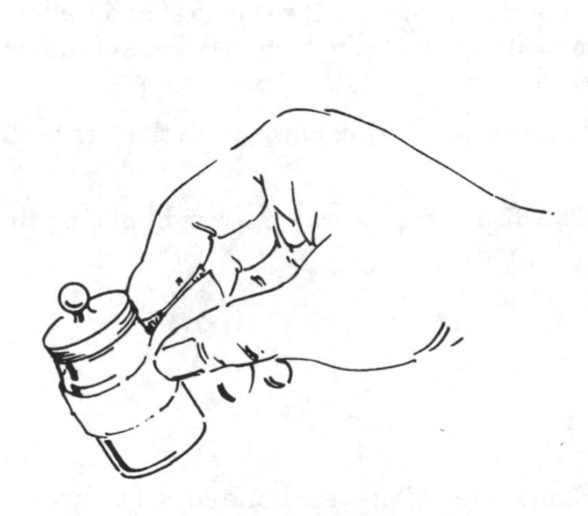

Handling a Weighing Bottle with a Paper Band

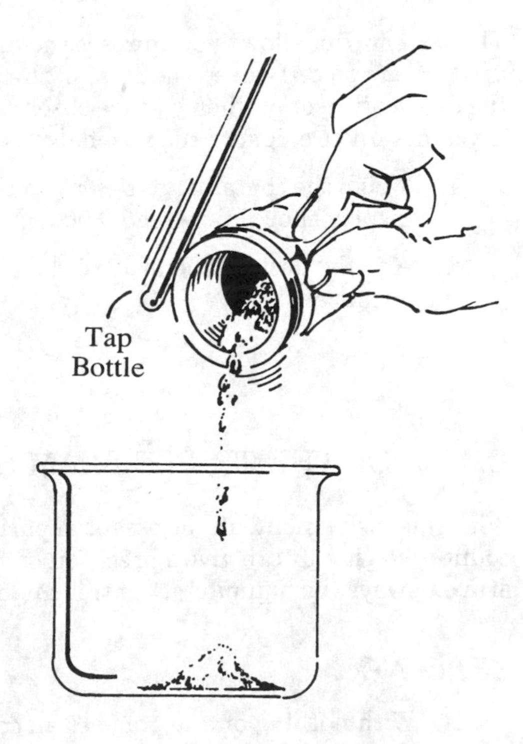

Transfer Procedure

II. Using Stock Reagent Bottles

All the chemicals you will use will be available in stock reagent bottles on the laboratory reagent shelf. The standard procedures for getting what you need from these bottles are described below. These procedures are to be used for two reasons: (1) To prevent contamination of the chemicals. Impure chemicals can give improper experimental results. (2) To prevent excessive wasting of chemicals. Most of the reagents are quite expensive, thus overuse of them is economically unsound.

In dealing with solid chemicals be sure the contents of the bottle are loose and in small pieces before you make a transfer. Often the chemical can be loosened by tapping the **closed** bottle against the palm of your hand. If this doesn't work, consult your instructor. Three methods of transferring solids are shown in the figures. The "hollow stopper" method and the direct pouring method are preferred. If you must use a spatula to scoop out the chemical be sure it is clean and dry. Do not go from one reagent bottle to the next without thoroughly cleaning and drying the spatula.

The Hollow Stopper Method

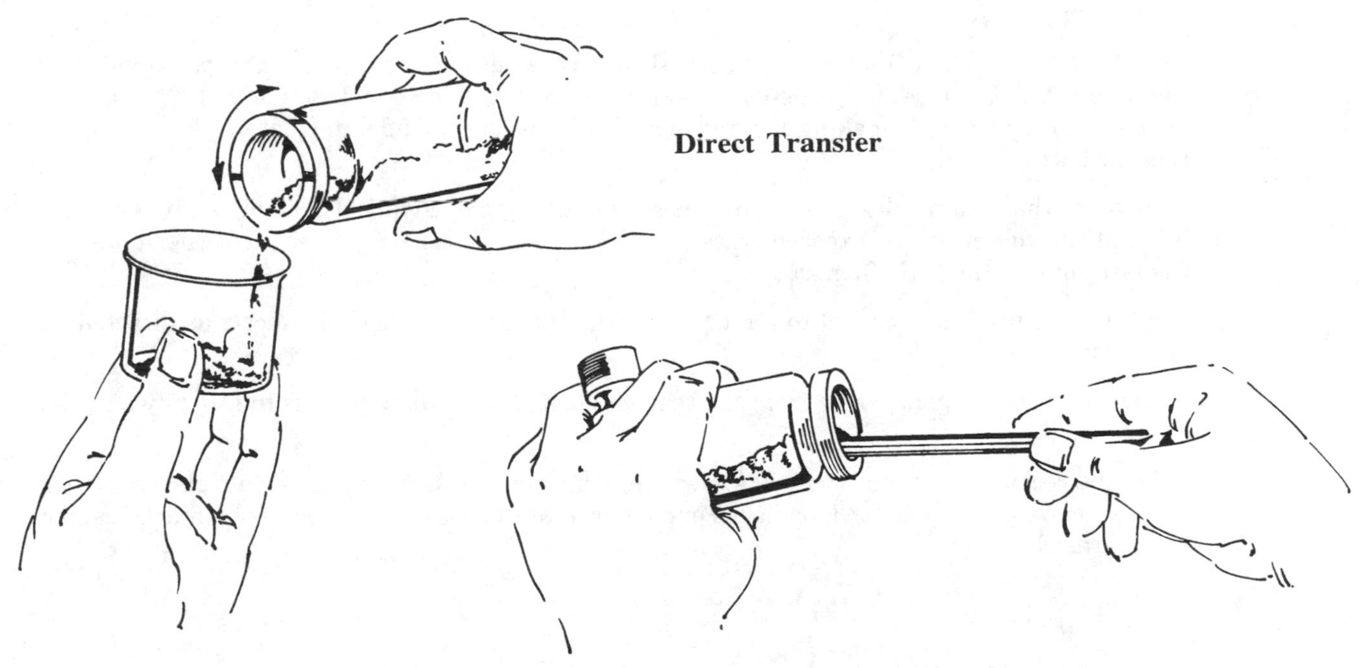

Direct Transfer

May Lead to Contamination

Transferring liquids is much simpler because of their fluidity. The preferred technique is shown in the figure, gentle pouring with the lip of the reagent bottle resting on the lip of the beaker. This technique prevents splashing and allows you to control the amount transferred. When you are pouring, hold onto the beaker to prevent it from tipping. If you need both hands to control the reagent bottle, have your partner hold the beaker. To prevent spills always use a receptacle whose opening is much larger than that of the reagent bottle. Thus you should not pour liquids from a reagent bottle directly into a test tube, nor should you pour from a one gallon reagent bottle into a 50 mL beaker.

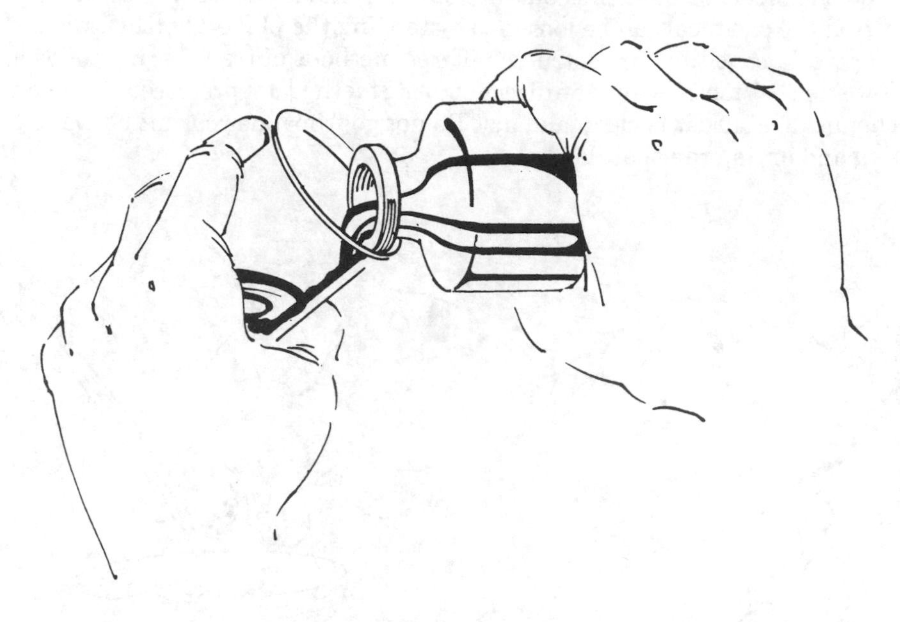

Liquid Transfer

Other rules that should be followed are:

1. Wipe up any spills immediately. If liquid runs down the side of a reagent bottle, wipe the bottle with a damp cloth.
2. Do not contaminate the reagent bottle lids. Handle the lids only by the finger grips provided. If the lid has a flat top, set it *upside-down* on the bench. If the top is not flat, do not set it on the bench, hold the lid while making the transfer. Always replace the lid immediately after using the reagent bottle.
3. Take only what you need. A useful guide is to assume 1 gm of chemical is about 1 cm^3. One cm^3 is about the volume of the exposed eraser on a new pencil. (Also 1 cm^3 is about ¼ the volume of the last joint on the little finger.)
4. Don't return unused chemical to a reagent bottle. Dispose of unused chemicals as directed by your instructor.
5. Never remove a reagent bottle from the reagent shelf. This is discourteous to others who need the chemical.
6. Label the receptacle so that you will know later which chemical it contains. Most glassware has white, etched glass spots for labelling. Pencil marks work best on these since the marks can be easily erased.

III. Burner Heating

The principle sources of heat for most experiments are gas burners that use natural gas (primarily methane, CH_4) and oxygen from the air to produce a hot, clean flame. When used properly the burner gives a flame whose temperature approaches 1500 °C. One of the simpler types of burners, called a Bunsen burner, is shown below.

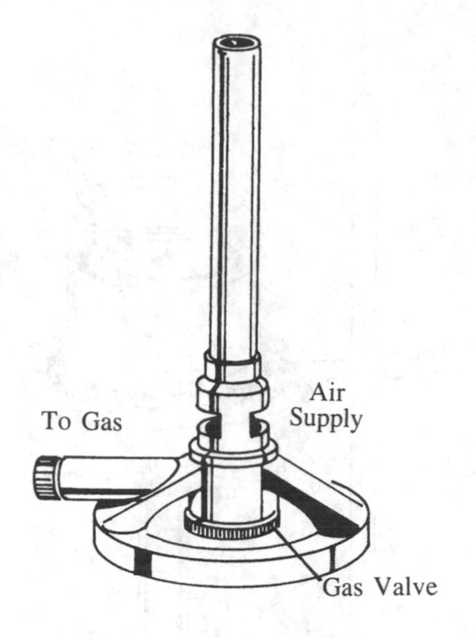

Bunsen Burner

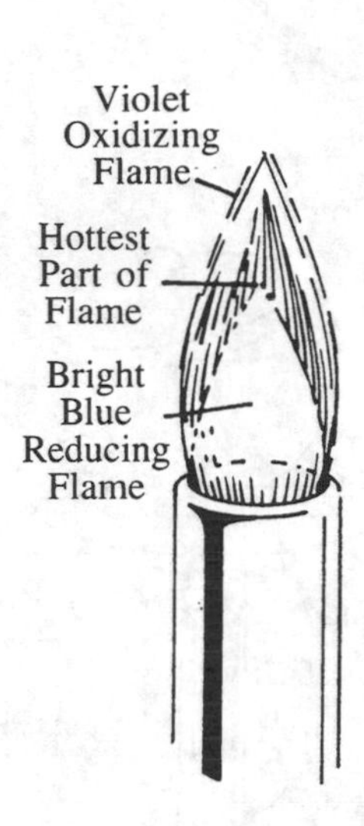

Two-Coned Flame

The following procedure is suggested for the first-time lighting of a particular burner.

1. Rotate the barrel to close the air holes. Shut the gas valve on the bottom of the burner, then open the valve by ¼ turn. Connect the tubing to the gas outlet on the bench.
2. Light a match and hold it by the barrel of the burner so that the flame licks over the rim of the barrel. Turn the gas valve of the bench outlet on full. This will produce a "lazy" flame.
3. Open the air holes by rotating the barrel until a sharp, two-coned flame is obtained. Adjust the gas valve on the burner to get a two-coned flame that is 1" to 2" high.

If the settings on the burner are not altered, subsequent lighting of the burner can be done by just holding a match flame to the barrel tip and turning on the bench valve.

The sharp, two-coned flame should always be used as this is the most consistent. To heat an object gently, move the flame in and out of position or place the object a few inches above the flame. Do not attempt to use a "lazy flame." For heavy heating, place the object at the tip of the center cone as this is the hottest part of the flame.

Solid chemicals can be burner heated in porcelain containers (such as crucibles or evaporating dishes) or in Pyrex type test tubes. Do not burner heat a solid in a beaker or flask as cracking of the glassware can occur. When heating liquids in a beaker or flask, it is suggested that you insert a stirring rod or boiling beads to prevent large gas bubbles from forming. **Do not** dry glassware using the burner. The heating is too intense and uneven and can result in shattering of the glassware. Use a drying oven or a clean towel if your glassware must be dried.

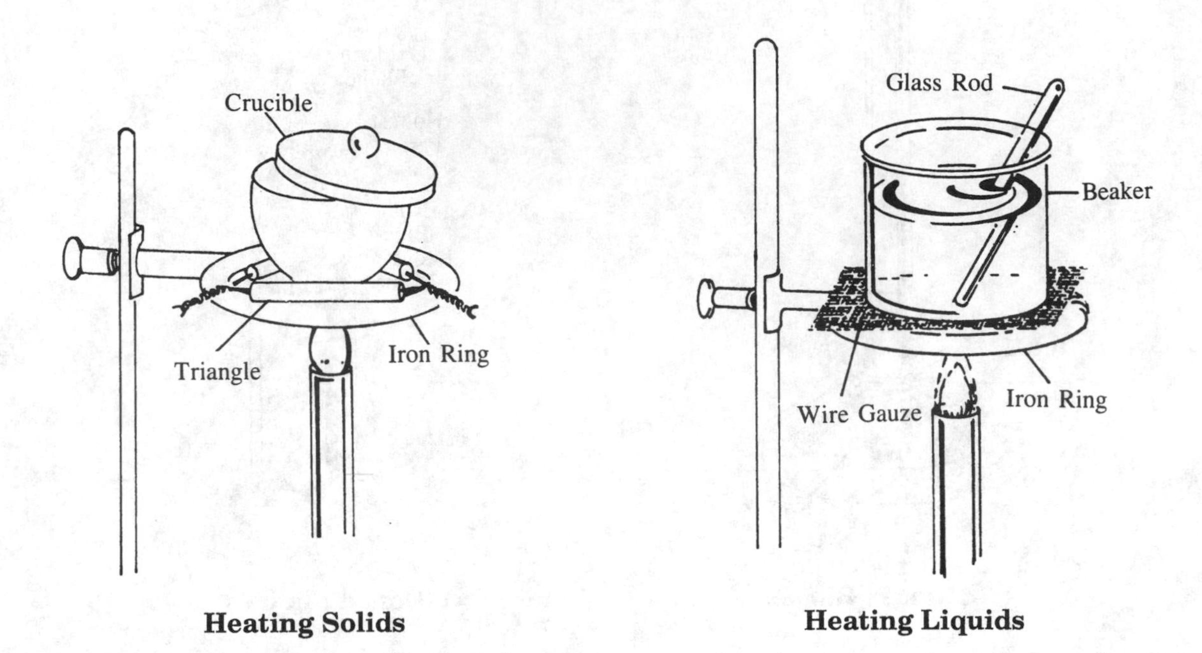

Heating Solids

Heating Liquids

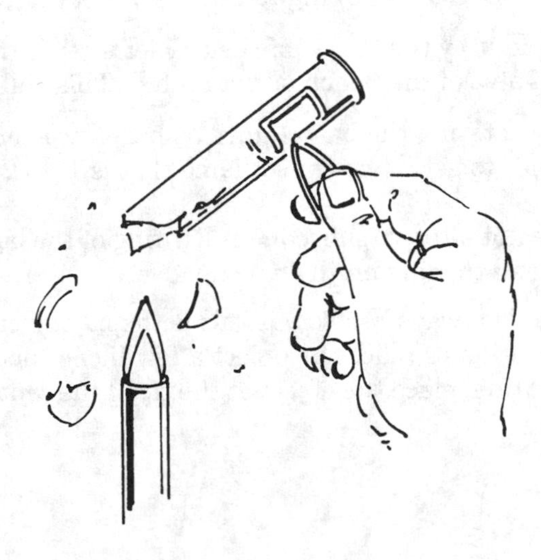

Do Not Dry Glassware with a Burner

IV. Manipulation of Glass Tubing and Rods

In some of your experiments you may need to fashion glass tubing into a certain size and shape or you may need to make stirring rods. The following procedures should be followed in such work.

A. BREAKING GLASS TUBING OR RODS

Using a triangular file or a metal scorer, make a single sharp scratch on the tubing at the point where the break is needed. Moisten the scratch. Grasp the tubing with the scratch facing away from you and with the thumbs located behind and on either side of the scratch. Apply a quick outward breaking force with the thumbs at the same time pulling away from the center (see figure below). The glass should break cleanly.

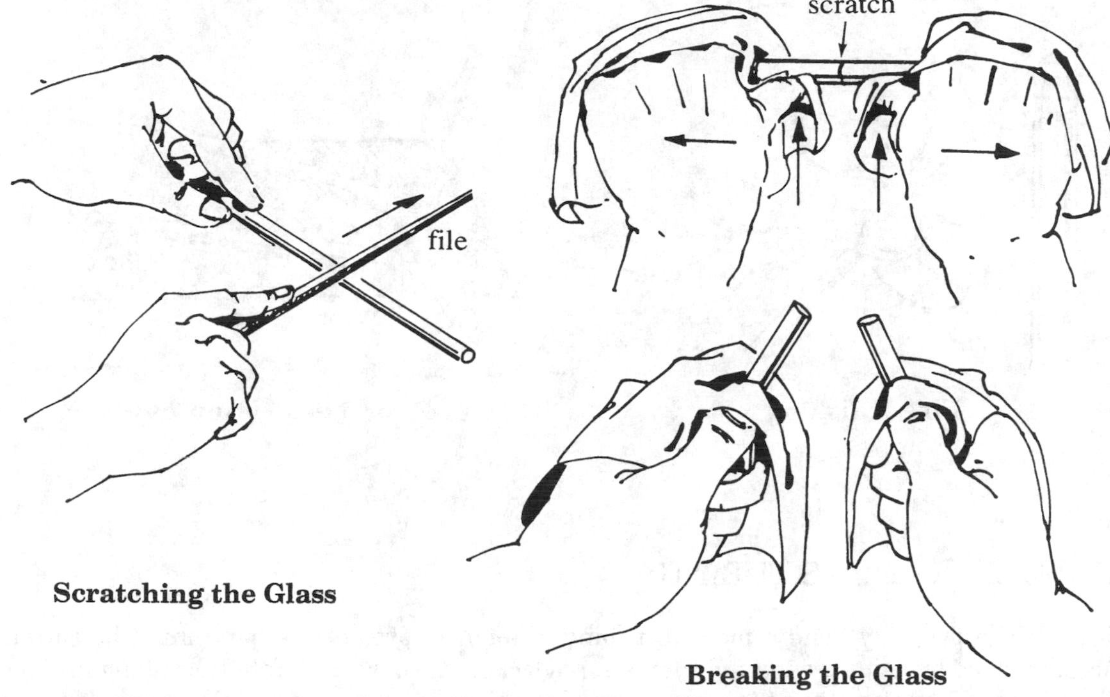

Scratching the Glass

Breaking the Glass

If you have trouble getting clean breaks, make the scratch a bit deeper. Also your results will improve with practice. Small spurs of glass can be smoothed by stroking them with a wire gauze.

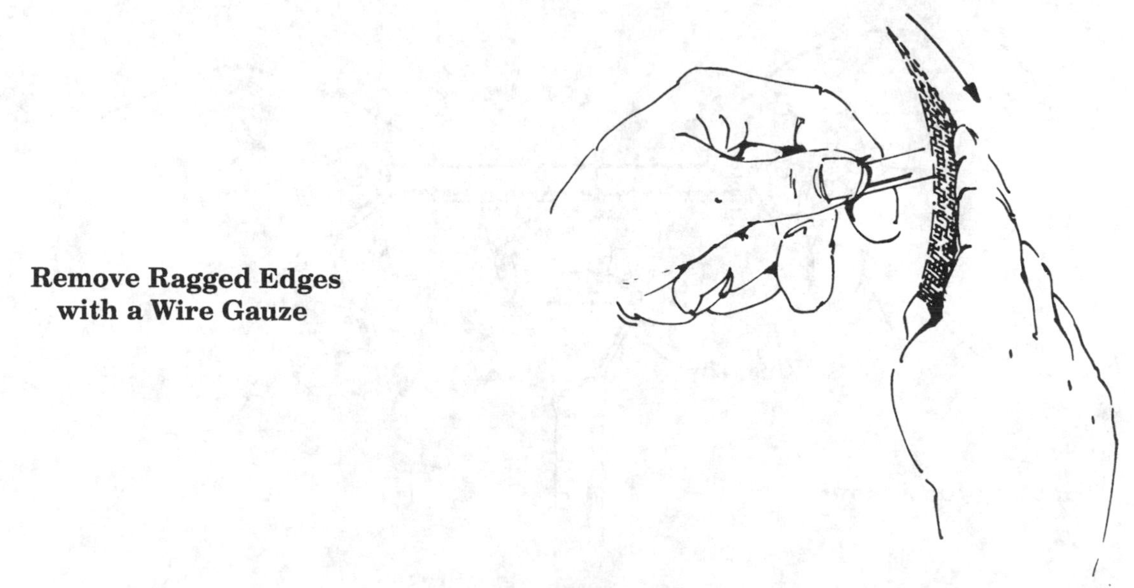

Remove Ragged Edges with a Wire Gauze

B. FIRE-POLISHING

The edge of a cut piece of tubing is very sharp and should be smoothed before use. Place the edge to be polished at the tip of the center cone of a burner flame and rotate the rod to insure even heating. Continue this action until the flame shows a yellow color (caused by Na from the glass being vaporized and ionized). At this point the glass should have softened enough to form a smooth edge. If the edge still appears sharp, continue heating. Avoid excessive fire-polishing as this can lead to a narrowing of the tube opening.

Place the hot piece of glass on a wire gauze to cool. Allow it to cool for 5 to 10 minutes then test the temperature of the heated end by gingerly touching it with a wet finger. **CAUTION!!!** Hot glass looks just like cold glass. Severe burns will result from handling uncooled glass.

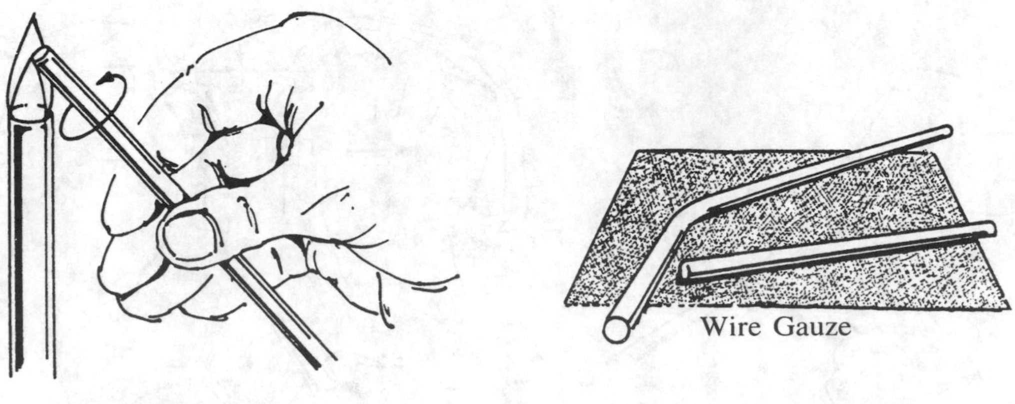

Fire Polishing **Cool for 5 to 10 Minutes**

C. BENDING GLASS TUBING

In order to properly bend a piece of tubing, the entire area of the bend must be softened. This is accomplished by using a wing-top (flame spreader) on the burner. Rotate the tubing in the high part of the wing-tip flame until the flame yellows and the glass becomes quite soft. Remove the tubing from the flame and smoothly bend it to the desired angle. (Too quick a bending will cause elongation and distortion of the glass.) Hold the bend in place for 10 to 15 seconds to allow it to harden. Place the bent tubing on a wire gauze to cool as described under the section on Fire-polishing.

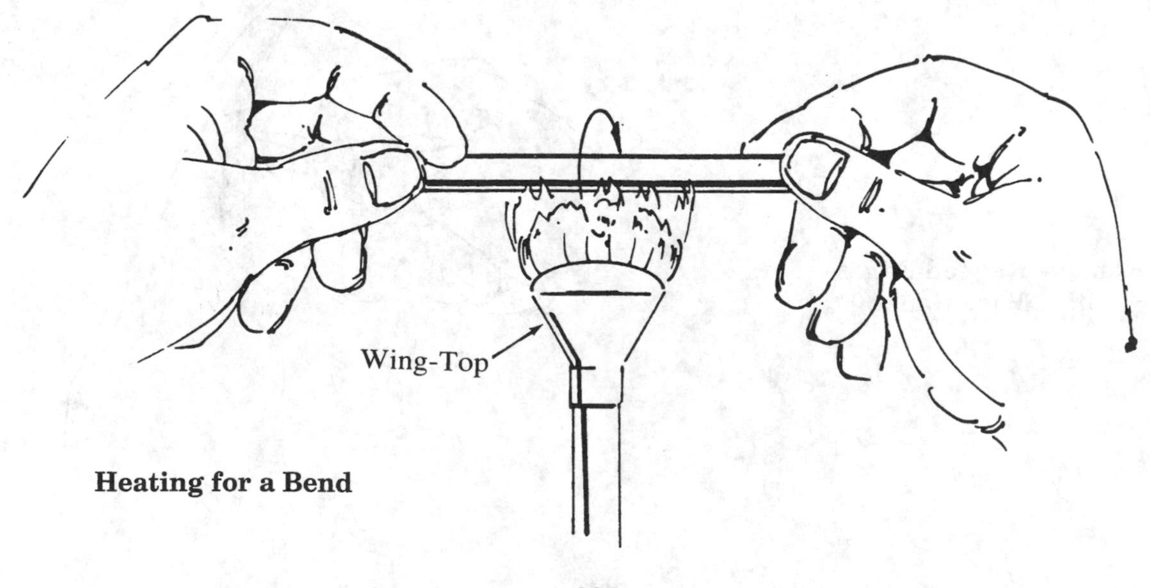

Heating for a Bend

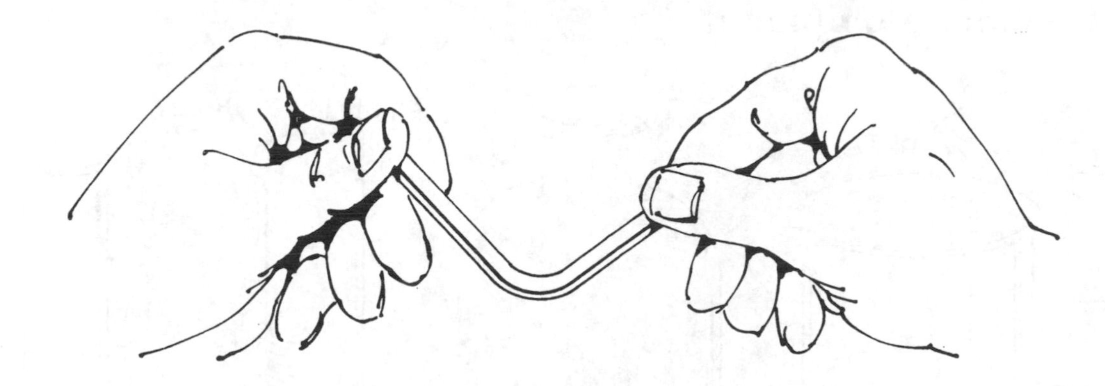

Bend Smoothly and Hold until Solid

D. INSERTING TUBING INTO A STOPPER

Inserting any glass stem (tubing, rods, thermometers) into the hole of a stopper is a potentially dangerous operation. The most common accident is for the glass to snap and be driven into the hands of the experimenter. Thus the task demands some skill and strict attention to safety precautions.

The tube to be inserted should have rounded edges (be fire-polished). The hole in the stopper should be just slightly smaller than the outer diameter of the tubing. Always insert from the side of the stopper that is to have the longest projection of tubing.

Wet the tip of the tubing and the stopper hole with water. Using toweling to protect ***both*** hands, grasp the tubing with the fingers only, about ***one inch*** from the end to be inserted. Insert by using a twisting motion with only moderate pushing. Continue this action, periodically lubricating with water, until the desired length of tubing is inserted.

NOTE: The toweling will protect both hands from receiving deep cuts if the tube breaks. By holding the tubing near the end being inserted the tension on the rod is minimized. Be careful that the palm of the hand does not push against the glass as this can snap it. If the tubing is not inserting smoothly, ***do not force it***—consult with the instructor.

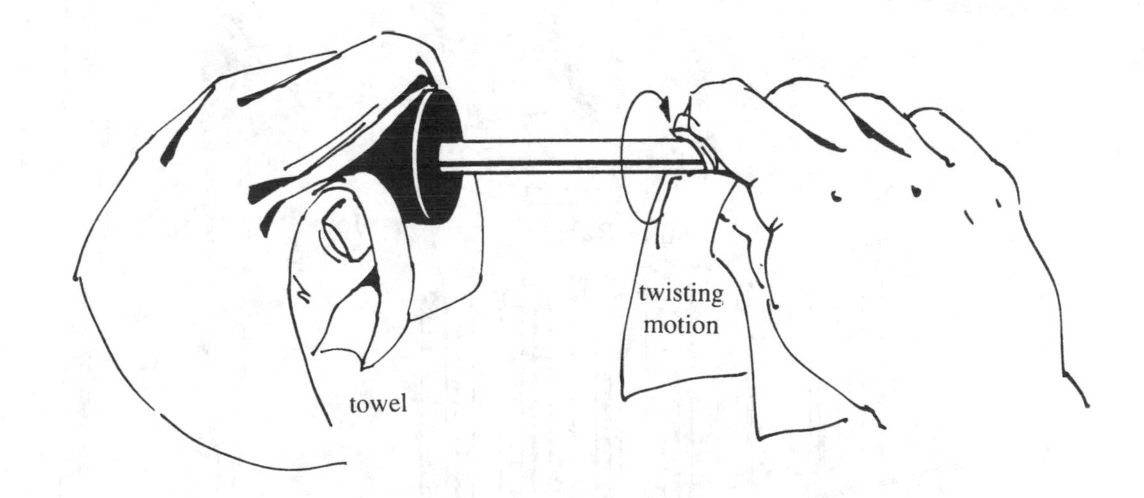

Inserting into a Stopper

Other Common Equipment

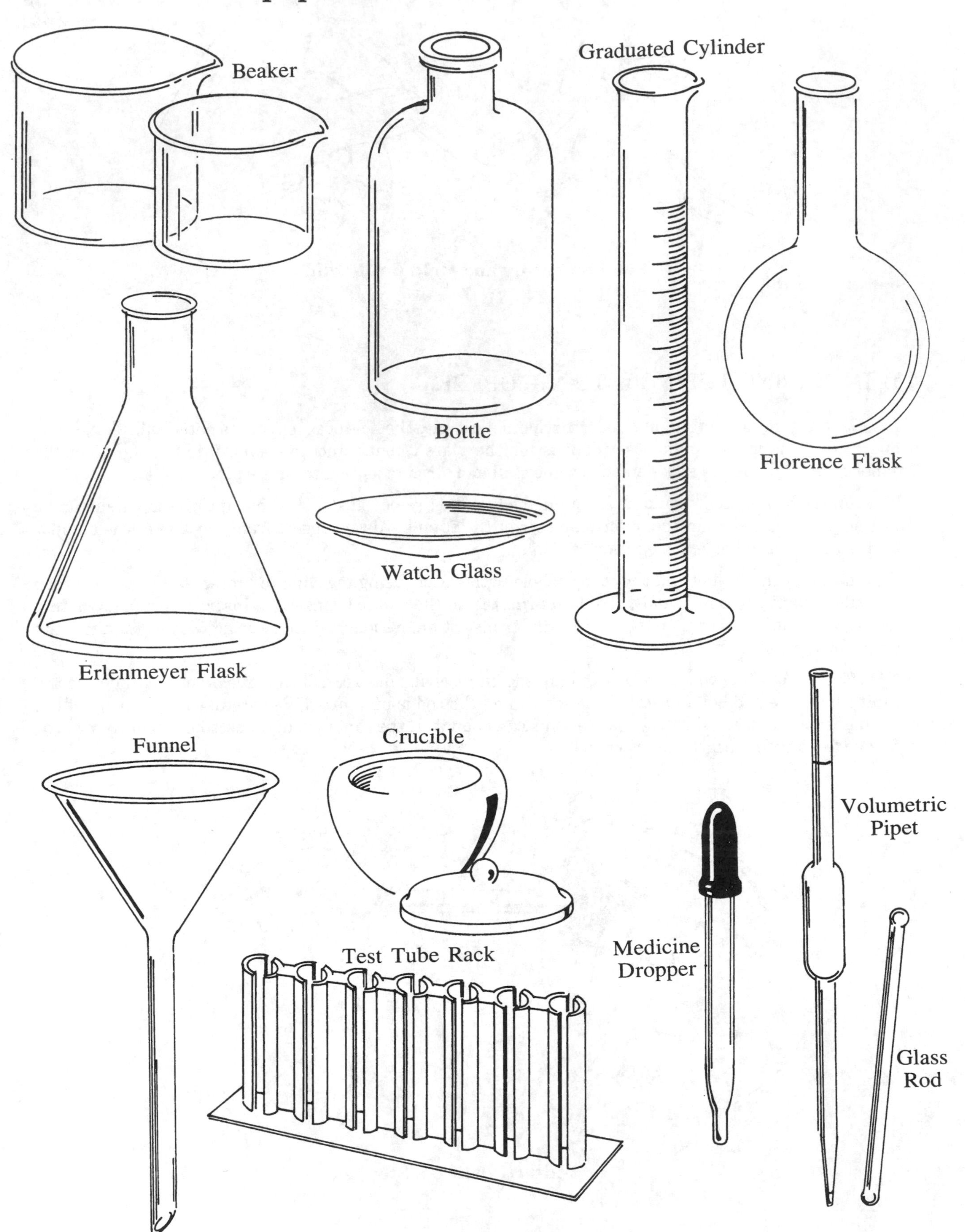

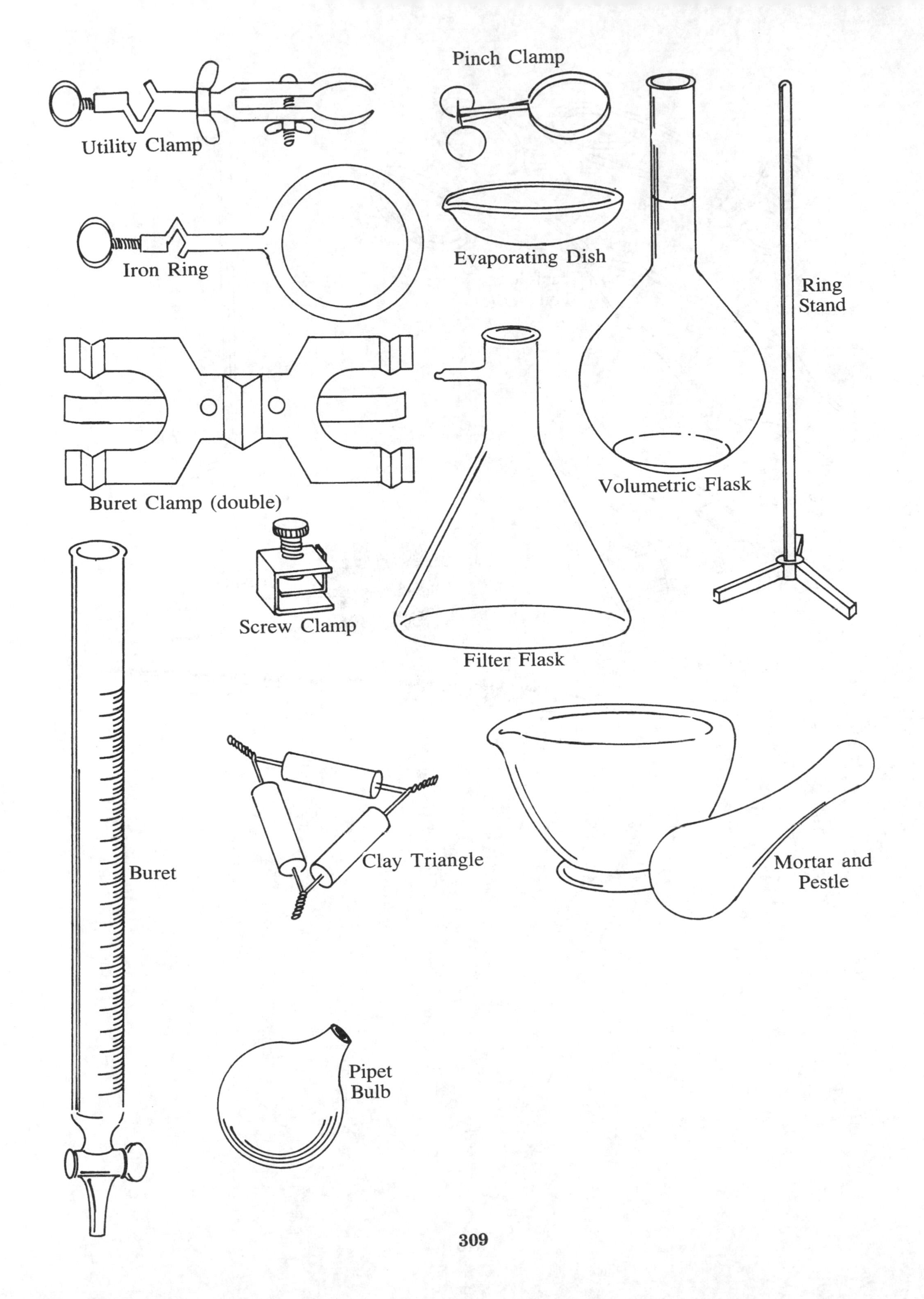
Pinch Clamp
Utility Clamp
Iron Ring
Evaporating Dish
Ring Stand
Buret Clamp (double)
Volumetric Flask
Screw Clamp
Filter Flask
Buret
Clay Triangle
Mortar and Pestle
Pipet Bulb

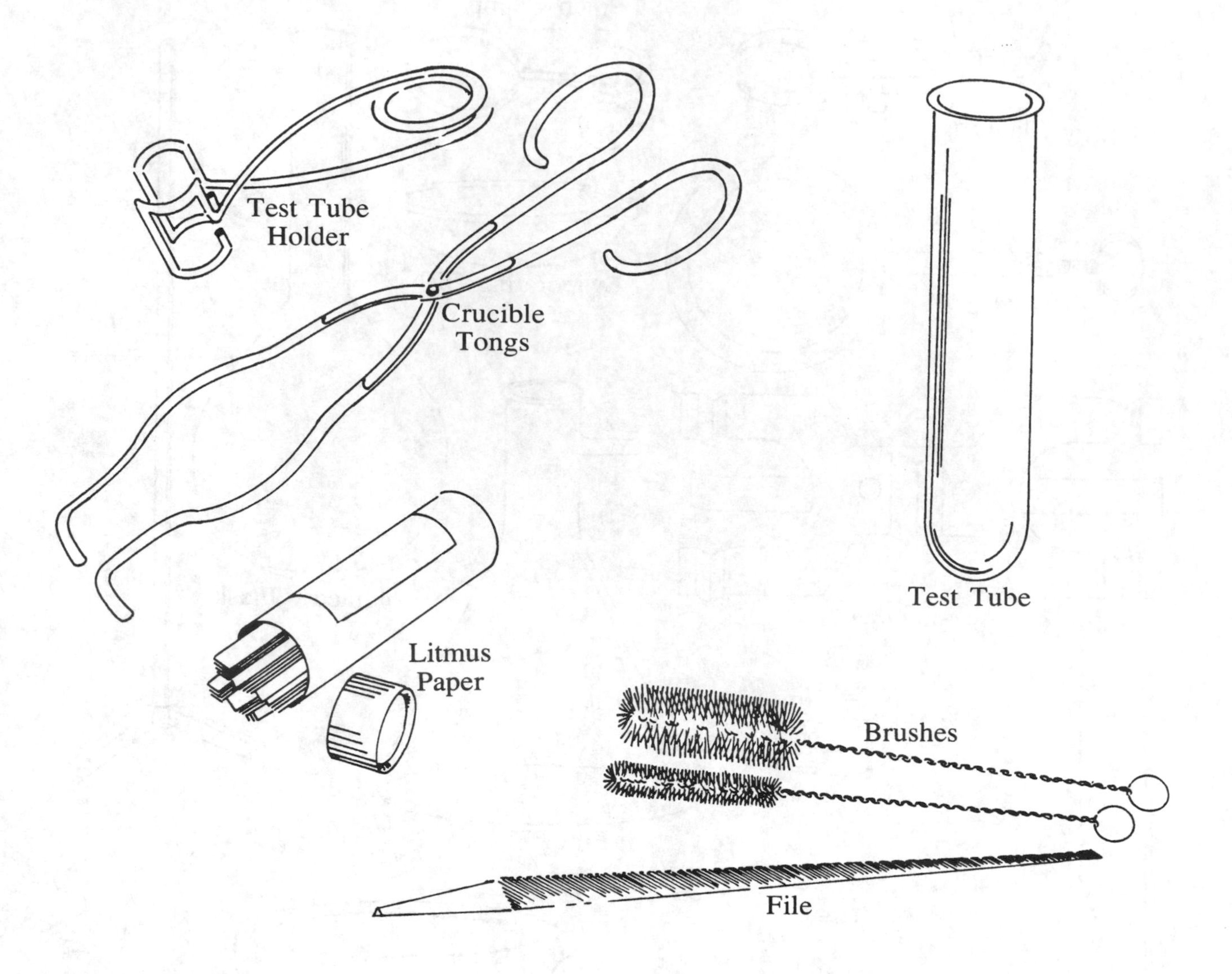
Test Tube
Holder
Crucible
Tongs
Test Tube
Litmus
Paper
Brushes
File

APPENDIX C

TABLES OF DATA

The following tables are necessarily abbreviated and incomplete. Readers are referred to more complete listings such as the Chemical Rubber Company's *Handbook of Chemistry and Physics.*

Table C-1: Physical Constants and Conversion Factors

1 atmosphere pressure (atm) = 760 mm of mercury(torr)

1 ampere current (amp) = 1 coulomb/sec:

1 Faraday = 96,500 coulombs, the amount of electricity which will deposit one equivalent of a substance at an electrode

$$R \text{ (gas constant)} = 0.08206 \ \frac{\text{L-atm}}{^\circ \text{K-mole}}$$

$$= 1.987 \ \frac{\text{cal}}{^\circ \text{K-mole}}$$

$$= 8.314 \ \frac{\text{J}}{^\circ \text{K-mole}}$$

Table C-2: Identification of Selected Chemicals

Products of chemical reactions can be identified by their chemical or physical properties. Following is a key to identifying some common chemicals.

Formula	Color	Odor	State at 25 °C	Solubility in H_2O	Miscellaneous
H_2	none	none	gas	no	burns explosively
O_2	none	none	gas	no	makes glowing split flame up
CO_2	none	none	gas	yes	turns lime water milky [saturated $Ca(OH)_2$]
NO_2	brown	choking	gas	yes	solution is acidic
I_2	purple	medicine	solid	yes	purple in hexane (or cyclohexane)
Br_2	red-brown	choking	liquid	yes	orange in hexane (or cyclohexane)
Cl_2	pale green	choking	gas	yes	yellow in hexane (or cyclohexane)
NH_3	none	choking (smelling salts)	gas	yes	solution is basic
SO_2	none	suffocating	gas	yes	solution is acidic
H_2S	none	choking	gas	yes	solution is neutral
HCl	none	choking	gas	yes	solution is acidic
H_2O	none	none	liquid	---	$CoCl_2$ paper turns pink

Table C-3: Solubility Rules

Following are general rules concerning the solubility of common ionic salts in water. These substances are considered soluble if they can form solutions of concentration greater than 0.1 M. They are considered partially soluble if they form solutions between 0.01 M and 0.1 M. They are considered insoluble if they form solutions with concentrations less than 0.01 M.

Ion	Rule	Important Exceptions
Na^+,K^+,NH_4^+	Salts of these ions are soluble.	Are no exceptions
NO_3^-,ClO_3^-, ClO_4^-, $C_2H_3O_2^-$	Salts of these ions are soluble.	$AgC_2H_3O_2$ is partially soluble.
Cl^-, Br^-, I^-	Salts of these ions are soluble.	Ag^+, Pb^{2+}, Hg_2^{2+} salts of these ions are insoluble; $PbCl_2$ is partially soluble.
SO_4^{2-}	Salts of this ion are soluble.	$BaSO_4$, $PbSO_4$, $SrSO_4$ are insoluble; $CaSO_4$, Hg_2SO_4, Ag_2SO_4 are partially soluble.
CO_3^{2-}, PO_4^{3-}, SO_3^{2-}	Salts of these ions are insoluble.	Na^+, K^+, NH_4^+ salts of these ions are soluble.
OH^-	Salts of this ion are insoluble.	$NaOH$, KOH, NH_4OH, and $Ba(OH)_2$ are soluble; $Ca(OH)_2$ is partially soluble.
S^{2-}	Salts of this ion are insoluble.	Na^+,K^+, NH_4^+, Mg^{2+}, Ca^{2+}, Sr^{2+}, and Ba^{2+} sulfides are soluble.

Using these rules it is possible to predict the formation of precipitates in chemical reactions. For example, in double displacement reactions (metathesis reactions) the solution of one ionic solid is added to the solution of a second ionic solid. If a precipitate is formed, use of solubility rules allows the identification of the precipitate. For example:

$Ba(NO_3)_2(aq)$	+	$Na_2SO_4(aq)$	$\longrightarrow$	$2NaNO_3(aq)$	+	$BaSO_4(s)$
nitrates are soluble		Na^+ salts are soluble, sulfates are soluble with exceptions		nitrates are soluble, Na^+ salts are soluble		sulfates are soluble, Ba^{2+} sulfate is an exception

Table C-4: Selected Acid-Base Indicators

Indicator	pH Range	Color Change (pH)
methyl orange	3.2–4.4	red to yellow
methyl red	4.8–6.0	red to yellow
bromothymol blue	6.0–7.6	colorless to yellow to blue
phenolphthalein	8.2–10.0	colorless to pink
alizarine yellow	10.1–12.0	yellow to red

Table C-5: Strength of Acids in Water at Room Temperature

Strength	Acid	Reaction	K_a*
strong	perchloric acid	$HClO_4 + H_2O \rightarrow H_3O^+ + ClO_4$	large
↑	hydrochloric acid	$HCl + H_2O \rightarrow H_3O^+ + Cl^-$	large
	nitric acid	$HNO_3 + H_2O \rightarrow H_3O^+ + NO_3^-$	large
	sulfuric acid	$H_2SO_4 + H_2O \rightarrow H_3O^+ + HSO_4^-$	large
	oxalic acid	$H_2C_2O_4 + H_2O \rightarrow H_3O^+ + HC_2O_4^-$	6.5×10^{-2}
	hydrogen sulfate ion	$HSO_4^- + H_2O \rightarrow H_3O^+ + SO_4^{2-}$	1.3×10^{-2}
	phosphoric acid	$H_3PO_4 + H_2O \rightarrow H_3O^+ + H_2PO_4^-$	7.5×10^{-3}
	formic acid	$HCHO_2 + H_2O \rightarrow H_3O^+ + CHO_2^-$	1.7×10^{-4}
	hydrogen oxalate ion	$HC_2O_4^- + H_2O \rightarrow H_3O^+ + C_2O_4^{2-}$	6.1×10^{-5}
	ascorbic acid	$HC_6H_7O_6 + H_2O \rightarrow H_3O^+ + C_6H_7O_6^-$	7.9×10^{-5}
	acetic acid	$HC_2H_3O_2 + H_2O \rightarrow H_3O^+ + C_2H_3O_2^-$	1.8×10^{-5}
	carbonic acid	$H_2CO_3 + H_2O \rightarrow H_3O^+ + HCO_3^-$	4.3×10^{-7}
	dihydrogen phosphate ion	$H_2PO_4^- + H_2O \rightarrow H_3O^+ + HPO_4^{2-}$	6.3×10^{-8}
	boric acid	$H_3BO_3 + H_2O \rightarrow H_3O^+ + H_2BO_3^-$	7.3×10^{-10}
	ammonium ion	$NH_4^+ + H_2O \rightarrow H_3O^+ + NH_3$	5.7×10^{-10}
	bicarbonate ion	$HCO_3^- + H_2O \rightarrow H_3O^+ + CO_3^{2-}$	4.8×10^{-11}
↓	monohydrogen phosphate ion	$HPO_4^{2-} + H_2O \rightarrow H_3O^+ + PO_4^{3-}$	4.4×10^{-13}
weak	water	$2H_2O \rightarrow H_3O^+ + OH^-$	$K=[H^+][OH^-] = 1 \times 10^{-14}$

*for $HB + H_2O \rightarrow H_3O^+ + B^-$, $K_a = \frac{[H_3O^+][B^-]}{[HB]}$

Table C-6: Standard Reduction Potentials for Selected Half-Reactions

Half-Reaction	E° (volts)
$2e^- + F_2(g) \rightarrow 2F^-(aq)$	2.87
$5e^- + 8H^+(aq) + MnO_4^-(aq) \rightarrow Mn^{2+}(aq) + 4H_2O(l)$	1.51
$2e^- + Cl_2(g) \rightarrow 2Cl^-(aq)$	1.36
$6e^- + 14H^+(aq) + Cr_2O_7^{2-}(aq) \rightarrow 2Cr^{3+}(aq) + 7H_2O(l)$	1.33
$2e^- + 4H^+(aq) + MnO_2(s) \rightarrow Mn^{2+}(aq) + 2H_2O(l)$	1.23
$4e^- + 4H^+(aq) + O_2(g) \rightarrow 2H_2O(l)$	1.23
$2e^- + Br_2(l) \rightarrow 2Br^-(aq)$	1.09
$3e^- + 4H^+(aq) + NO_3^-(aq) \rightarrow NO(g) + 2H_2O(l)$	0.96
$e^- + Ag^+(aq) \rightarrow Ag(s)$	0.80
$e^- + Fe^{3+}(aq) \rightarrow Fe^{2+}(aq)$	0.77
$3e^- + 2H_2O(l) + MnO_4^-(aq) \rightarrow MnO_2(s) + 4OH^-(aq)$	0.59
$2e^- + I_2(s) \rightarrow 2I^-(aq)$	0.53
$4e^- + 2H_2O + O_2(g) \rightarrow 4OH^-(aq)$	0.40
$2e^- + Cu^{2+}(aq) \rightarrow Cu(s)$	0.34
$e^- + Cu^{2+}(aq) \rightarrow Cu^+(aq)$	0.15
$2e^- + Sn^{4+}(aq) \rightarrow Sn^{2+}(aq)$	0.13
$2e^- + 2H^+(aq) \rightarrow H_2(g)$	0.00
$3e^- + 4H_2O(l) + CrO_4^{2-}(aq) \rightarrow Cr(OH)_3(s) + 5OH^-(aq)$	-0.12
$2e^- + Pb^{2+}(aq) \rightarrow Pb(s)$	-0.13
$2e^- + Sn^{2+}(aq) \rightarrow Sn(s)$	-0.14
$2e^- + Ni^{2+}(aq) \rightarrow Ni(s)$	-0.23
$2e^- + Fe^{2+}(aq) \rightarrow Fe(s)$	-0.44
$2e^- + Zn^{2+}(aq) \rightarrow Zn(s)$	-0.76
$2e^- + 2H_2O(l) \rightarrow H_2(g) + 2OH^-(aq)$	-0.83
$3e^- + Cr(OH)_3(s) \rightarrow Cr(s) + 3OH^-(aq)$	-1.30
$3e^- + Al^{3+}(aq) \rightarrow Al(s)$	-1.66
$2e^- + Mg^{2+}(aq) \rightarrow Mg(s)$	-2.37
$e^- + Na^+(aq) \rightarrow Na(s)$	-2.71
$2e^- + Ca^{2+}(aq) \rightarrow Ca(s)$	-2.87
$e^- + K^+(aq) \rightarrow K(s)$	-2.93

Table C-7: Organic Compounds, Lewis Structures and Properties

Name	Lewis Structure**	Properties
Octyl acetate	$CH_3\text{-}C(=O)\text{—}O\text{—}CH_2CH_2CH_2CH_2CH_2CH_2CH_2CH_3$	odor of oranges boil pt. = 210 °C
Iso amyl-acetate	$CH_3\text{—}C(=O)\text{—}O\text{—}CH_2CH_2CH(CH_3)CH_3$	odor of bananas boil pt. = 142 °C
Ethyl butanoate	$CH_3CH_2CH_2\text{—}C(=O)\text{—}O\text{—}CH_2CH_3$	odor of pineapple boil pt. = 110 °C
Aniline	$C_6H_5\text{—}NH_2$ (benzene ring C—N—H, with H on N)	biting odor boil pt. = 184 °C
Acetanilide	$C_6H_5\text{—}N(H)\text{—}C(=O)\text{—}C\text{—}$	shiny, platelike crystals in water melt pt. = 114 °C
Benzamide	$C_6H_5\text{—}C(=O)\text{—}N(H)\text{—}H$	soluble in water melt pt. = 132 °C

**C— indicates a C-H bond

APPENDIX D

SPECIALIZED TECHNIQUES

D.1 Volume Measurements

There are several pieces of glassware designed to measure liquid volumes. Some give only approximate measurements (droppers, graduated beakers, and flasks) while others give very precise measurements (graduated cylinders, pipets, burets). When using these more precise measurers, the experimenter must be aware of two factors: how to read a meniscus and whether the glassware is designed to hold or to deliver the measured volume.

The liquid surface in any tube will be slightly curved, not perfectly horizontal. This curved liquid surface is called a ***meniscus***. It will be curved downwards (as with water) when the attraction of the liquid molecules to the atoms of the glass is stronger than the attraction the liquid molecules feel toward each other. With liquids such as mercury, the meniscus is curved upward because the Hg atoms feel a stronger attraction for each other.

When measuring the level of a liquid, we must decide which part of the meniscus to read. Since consistency is the important consideration, chemists read the rounded portion of the meniscus (i.e., the bottom of the meniscus for H_2O solutions). When reading a liquid level, the eye must be ***on the same horizontal level*** as the meniscus. This is to avoid the problem of parallax, choosing the wrong measuring line due to an improper viewing angle.

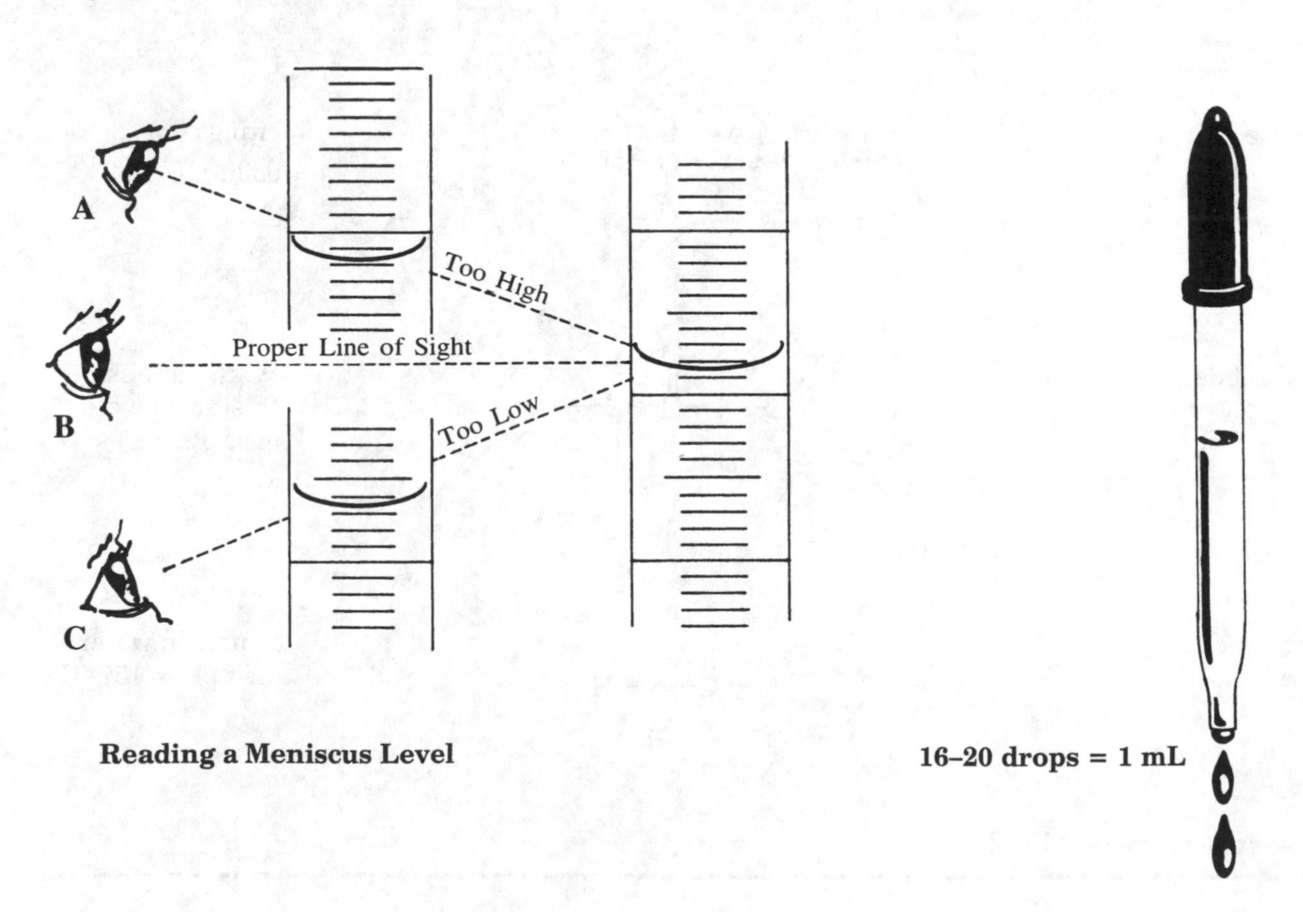

Reading a Meniscus Level

16–20 drops = 1 mL

The more precise measuring glassware is marked to measure either the amount of liquid being ***held*** by or the amount of liquid that can be ***delivered*** (extracted) from the glassware. No one volume marking can precisely measure both of these because all the liquid cannot be poured from a container. Some will always remain behind, clinging to the container walls. The experimenter should be aware of which glassware is marked to measure liquid held or liquid delivered and use this glassware accordingly.

A. DROPPERS AND GRADUATED BEAKERS

These are used to give approximate measures of volumes. The dropper is the quickest way to measure volumes of less than a few milliliters. The measurement is made by counting free falling drops taking 1.0 mL = 18 drops. The method is precise to ±10% as one mL may be equal to 16 to 20 drops depending on the particular dropper used. For larger volumes the markings on beakers and flasks can be used. These are accurate to only ±5% because of the large diameters of the glassware and the methods of manufacture. It is not important here whether the volume marking refers to the volume held or delivered since the inherent error is so large.

B. GRADUATED CYLINDERS

The volume markings here measure the amount of liquid held by the cylinders. Their accuracy is about ±1 to 2%. The error is much larger, of course, if the cylinder is used to measure liquid delivered. One way around this problem is to do some calibrations ahead of time. For example, assume you wish to use a graduated cylinder to deliver about 100 mL of solution several times in an experiment. Take two 100 mL cylinders and fill one with water to the 100.0 mL mark. Now pour this water gently into the second, dry, cylinder. Read the volume delivered. Within the error limits of ±1 to 2%, this is the volume the first cylinder will deliver throughout the experiment. (This assumes that your pouring procedure will be the same during the experiment.)

C. PIPETS AND BULBS

Pipets are used to deliver measured volumes with an uncertainty of only about ±.02% and should be used whenever highly accurate work is desired. There are two general types: volumetric pipets, which have only one volume mark, and graduated pipets, which have several marks. The pipetting bulb is used to draw the liquid into the pipet. (There is the temptation to disregard the bulb and draw the liquid into the pipet by mouth suction. Do not give in to temptation! Always use the bulb! Mouth suction can easily lead to ingestion of corrosive or poisonous chemicals.) The proper use of the pipet and bulb requires a fair amount of skill. It may be worth the time to practice and to perfect your skills using distilled water.

PREPARING THE PIPET

In order to not contaminate or dilute the solution being measured, the pipet must be cleaned and its inside coated with the solution to be measured.

1. Draw distilled water into the pipet until about 1/5 of its "fat" portion is filled. Take off the bulb, quickly placing your thumb or forefinger over the pipet end. This will keep the water from draining out. Keeping your finger in place, tip the pipet to a horizontal position. Rotate the pipet so that the entire "fat" portion is repeatedly coated with water.

   Now allow the water to move down the "bulb" stem until it is about 2/3 of the way between the volume mark and the end of the stem. This is done by tipping the pipet to about a 45-degree angle and slightly releasing your finger pressure. (It is wise to not let any liquid reach the end of the "bulb" stem. If the end is wet it will be almost impossible to control the flow of liquid with finger pressure during the actual measurement.)

   Tip the pipet back to the normal vertical position ("bulb" stem up) and let the water drain into a sink. Check the coating of the inside. If it is uniform, having no breaks or localized droplets formed, proceed to Step #3.

2. An uneven coating with water indicates that some insoluble chemicals, such as "oils," are still adhering to the glass. Wash the pipet 2 to 3 times with a mild soap solution using the procedure outlined in Step #1 above. Then rinse out the soap solution with 4 to 6 treatments of tap water.

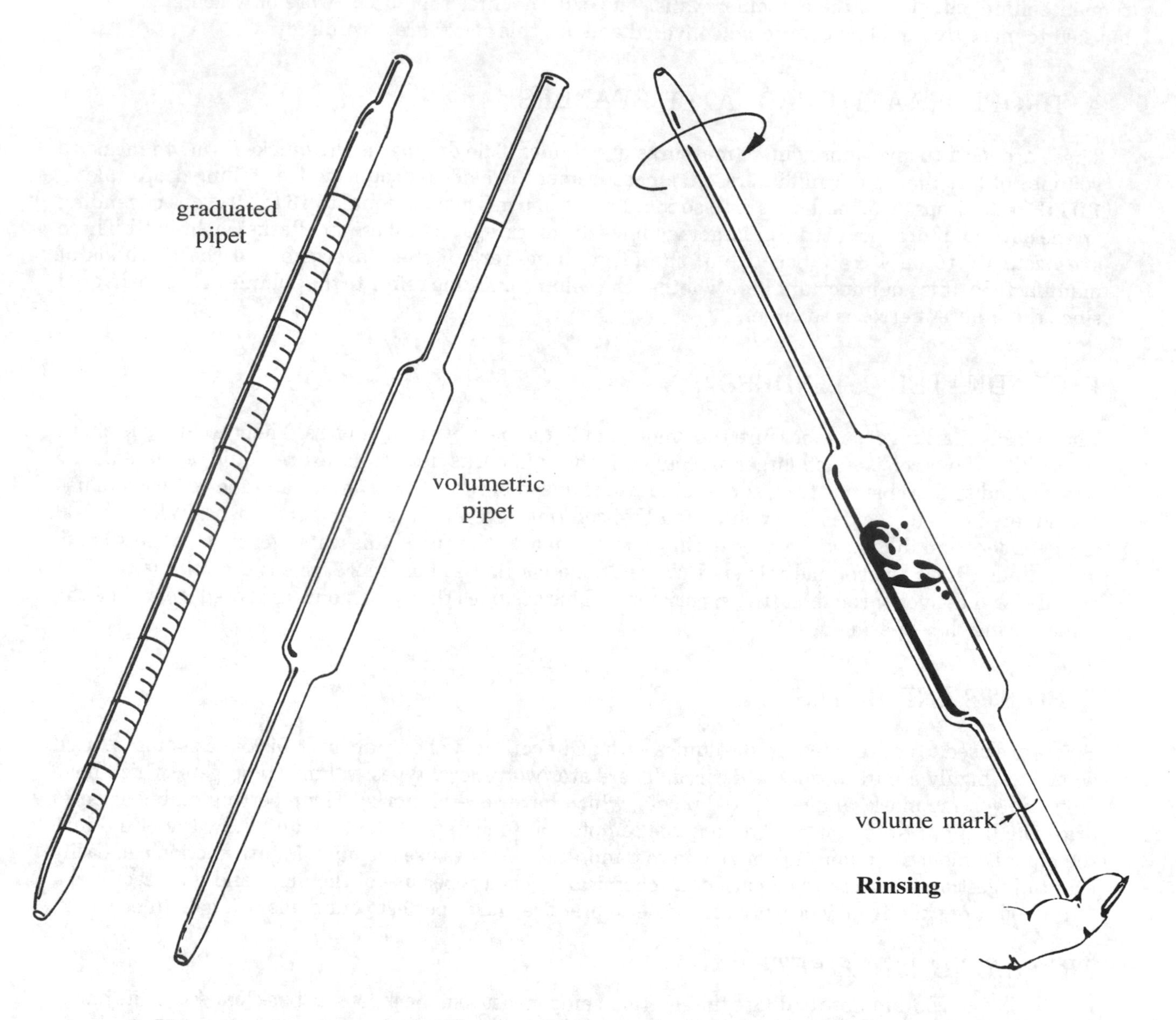

3. Using the procedure outlined in Step #1 above, rinse the pipet twice more with distilled water. Then rinse it twice with the solution to be measured. This will replace the water coating with a coating of the solution.

MEASURING PROCEDURE

The inside of the pipet should be coated with the solution to be measured as described above. If you will be using the same pipet to measure several solutions of the same chemical at different concentrations, it is advisable to start with the one of lowest concentration.

4. Attach the bulb to the pipet stem to a depth of no more than a few millimeters. (If the stem is inserted too deeply into the bulb, it will be difficult to separate the two later.) Compress the bulb and insert the tip of the pipet into the solution. Slowly release the pressure on the bulb, allowing the solution to be drawn up the pipet. Continue releasing the bulb pressure until the liquid level is about halfway between the volume mark and the bulb.

5. Keeping the pressure on the bulb constant, push the bulb off the stem with the thumb or forefinger of the other hand. In the same action, quickly clamp the thumb or forefinger over the stem end. This will hold the liquid level above the volume mark.

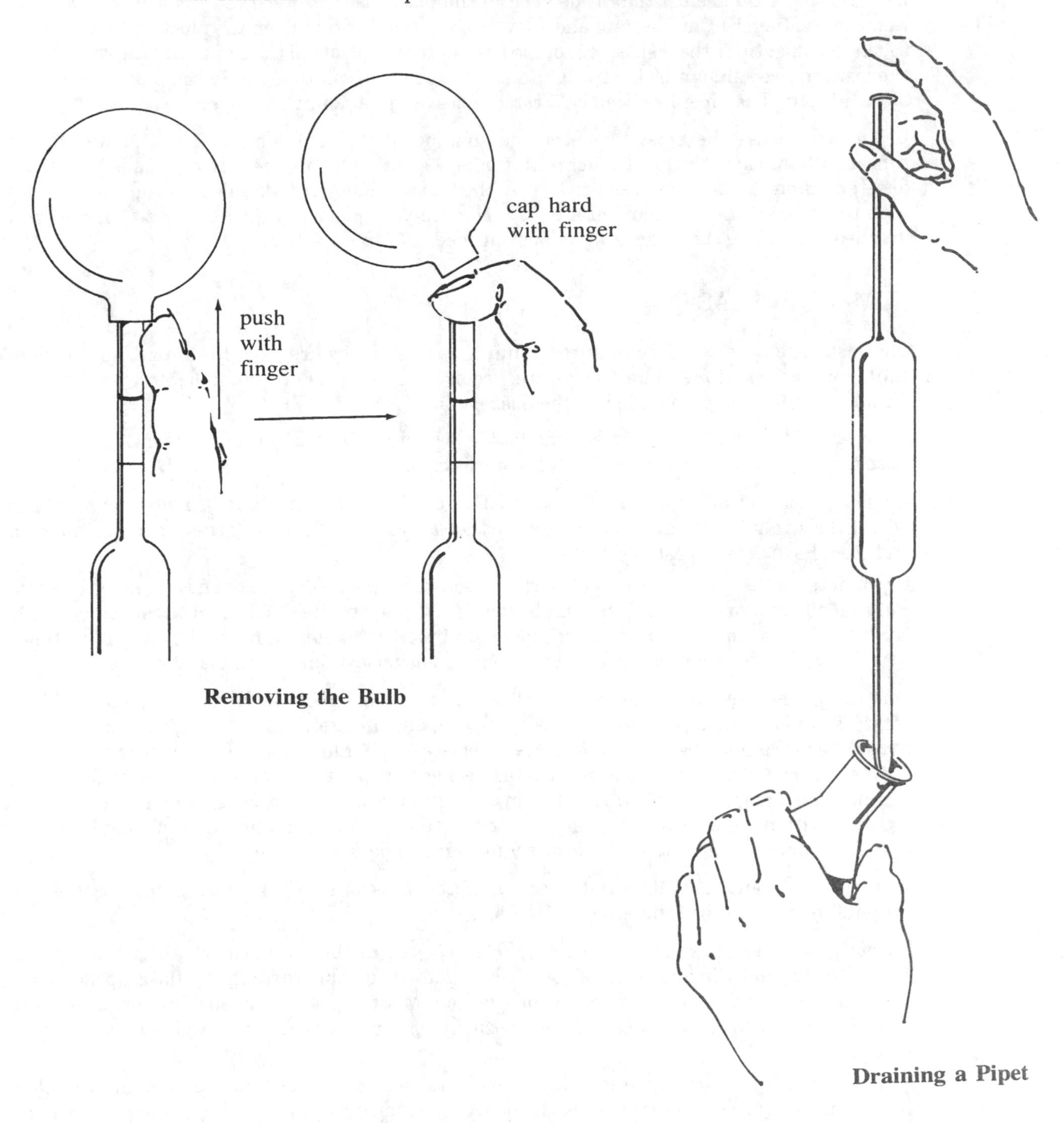

Removing the Bulb

Draining a Pipet

6. Wipe the pipet tip with a clean towel. Release your finger pressure ever so slightly, allowing the solution to drain ***just to*** the volume mark. This can be done by slightly rotating the finger on the stem or by slowly relaxing the tension in the finger. The draining solution can be caught in the beaker holding the remainder of the solution. Touch the tip of the pipet to the beaker to remove any partially formed drops.

If the liquid level falls below the volume mark or if any air bubbles are present in the pipet, drain the solution back into the original beaker and start over.

7. Hold the pipet tip against the inside glass of the flask that is to receive the measured volume. Remove the finger from the stem and allow the solution to drain from the pipet. After drainage, touch the pipet tip to the inside glass of the flask or to the surface of the solution. This will remove the last required amount. There will be a small amount of solution still in the pipet tip; ***do not*** blow this out. The pipet has been calibrated to leave this amount undelivered.
8. If you are to use the pipet to measure a solution of the same chemical but at a different concentration, rinse the pipet twice with the new solution before measuring. If a solution of a different chemical is to be measured, rinse the pipet 3 times with distilled water, then 2 times with the new solution. When you are through with your pipet work, rinse it 3 to 5 times with distilled water. Use the rinsing directions outlined in Step #1 above.

D. VOLUMETRIC FLASKS

Volumetric flasks are used to hold measured volumes. Usually they are used to prepare solutions of known molar concentrations. The flasks are precise to $\pm.02\%$. One must be concerned with contamination as well as precision in using the flasks.

In the following instructions, it is assumed that distilled water is to be the solution solvent. If this is not the case, use the desired solvent where distilled water is called for.

1. Rinse the flask 3 times with distilled water. This can be done by putting a small amount (about 1/5 the flask volume) of water in the flask and then rotating the flask until the entire "fat" portion and stem have been repeatedly coated.

   If you note that the water coating of the flask is uneven, the flask is dirty and should be cleaned. Put a small amount of soap solution in the flask. Shake and rotate the flask until the entire inside is coated. Repeat this treatment twice more with fresh soap solution. Rinse the flask 4 to 6 times with tap water to remove all traces of the soap. Then rinse 3 times with distilled water.
2. Quantitatively (precisely) transfer the solute (the material to be dissolved) into the flask. If the solute is liquid, measure it into the flask using a pipet or a buret. If it is a solid, accurately weigh the solute into a small beaker or onto a creased piece of weighing paper. Hold the beaker or paper against the rim of the flask opening. Tip the beaker or paper to a slight angle and use a spatula to scoop small amounts of solid at a time into the flask. Finally rinse the spatula and the beaker or paper with distilled water, catching ***all*** the rinse in the flask. These things should be rinsed at least 3 times to insure that all of the solute is in the flask.
3. Add distilled water until the "fat" portion of the flask is about 4/5 full. Swirl the flask to thoroughly dissolve the solute.
4. Carefully add distilled water until the liquid level is ***just at*** the volume mark. Stopper the flask. Mix the solution by inverting the flask. This is done by quickly turning the flask upside down allowing the air gap to move through the solution. Mix by inversion about 7 times. Be sure to hold the stopper firmly in place. (Mixing by shaking simply does not work well with volumetric flasks.)

   After mixing and removing the stopper, you will often notice that the liquid level is slightly below the volume mark. This is simply because some of the solution is now coating the upper stem and stopper. Do not be concerned and ***do not*** add more distilled water.

E. BURETS

Burets are calibrated to measure the volume of liquid delivered and are precise to $\pm.02\%$. Burets are used whenever odd volumes are to be delivered or, more commonly, in titration analysis experiments.

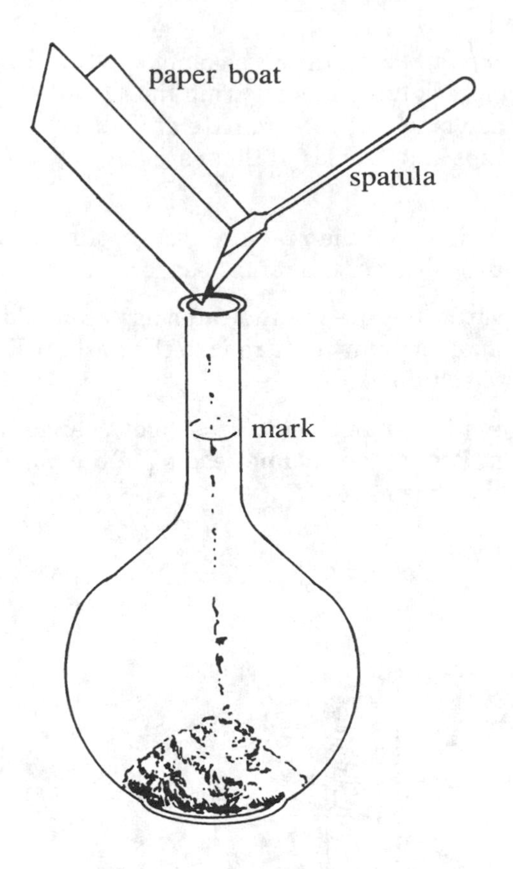

Adding a Solid to a Volumetric Flask

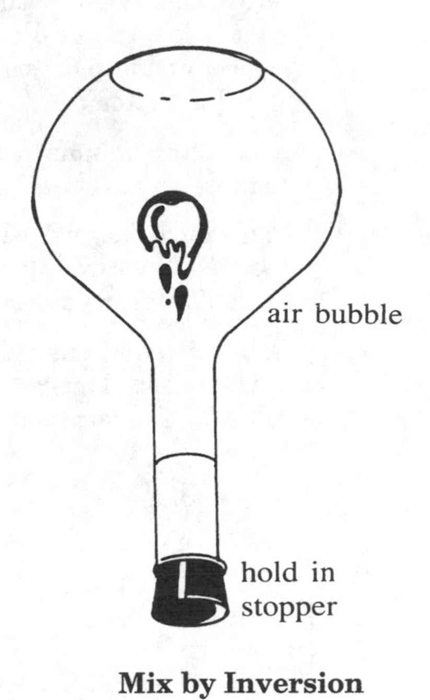

Mix by Inversion

The purpose of a titration is to determine the precise volume of a solution needed for a reaction. This is done by using the buret to add small amounts of the solution to the reaction flask until some indicator changes color. The amount of solution added is determined by noting the buret volume readings before and after these additions. The solution in the buret is called the ***titrant***. The point in the experiment where the indicator permanently changes to the desired color is called the ***end-point.***

PREPARING THE BURET

In order not to contaminate or dilute the solution being measured (the titrant), the buret must be cleaned and its inside coated with the titrant before use.

1. Rinse the buret 3 times with distilled water. This can be done by filling the buret 1/5 full with distilled water and then tipping and rotating it to coat the entire inside with the water. Next hold the buret upright and turn the stopcock, allowing the water to drain into a "refuse" beaker. Note the coating of water on the buret walls.
2. If the water coating is uneven, showing breaks or droplets, the buret is dirty and should be cleaned with soap solution. Add some soap solution and clean the entire barrel using a buret brush. If brushes are not available, rinse the buret 3 times with soap solution using the procedure outlined in Step #1 above. Be sure to drain some soap solution through the stopcock and tip.

   Rinse the buret barrel, stopcock and tip about 5 times with tap water to remove all traces of soap. Then rinse these parts 3 times with distilled water.
3. Using the procedure in Step #1, rinse the buret barrel, stopcock and tip at least twice with the titrant solution. This will replace the water coating with a coating of titrant.

TITRATION PROCEDURE

4. Secure the buret to a stand. Fill it with titrant to a point above the zero volume mark. Place the refuse beaker under the buret and turn on the stopcock full force, allowing the titrant to drain a bit below the zero mark. This full-force drainage is meant to eliminate any air gaps in the stopcock and tip. Check the tip and barrel for air gaps and bubbles. If there are any, consult with your instructor.
5. Record the volume reading on the buret. (Your initial volume reading need not be at zero.) Remove any drops on the tip by touching the tip to the side of the refuse beaker.
6. Place the solution flask to be titrated under the buret. Be sure the indicator has been added to the flask. By carefully turning the stopcock, add small amounts of titrant to the flask. Swirl the flask after each addition to completely mix the two solutions.
7. Continue additions until one drop of titrant causes a permanent (30 sec.) color change in the swirled flask. This is the end-point of the titration. Record the volume reading of the buret and calculate the total volume of titrant solution that has been added.

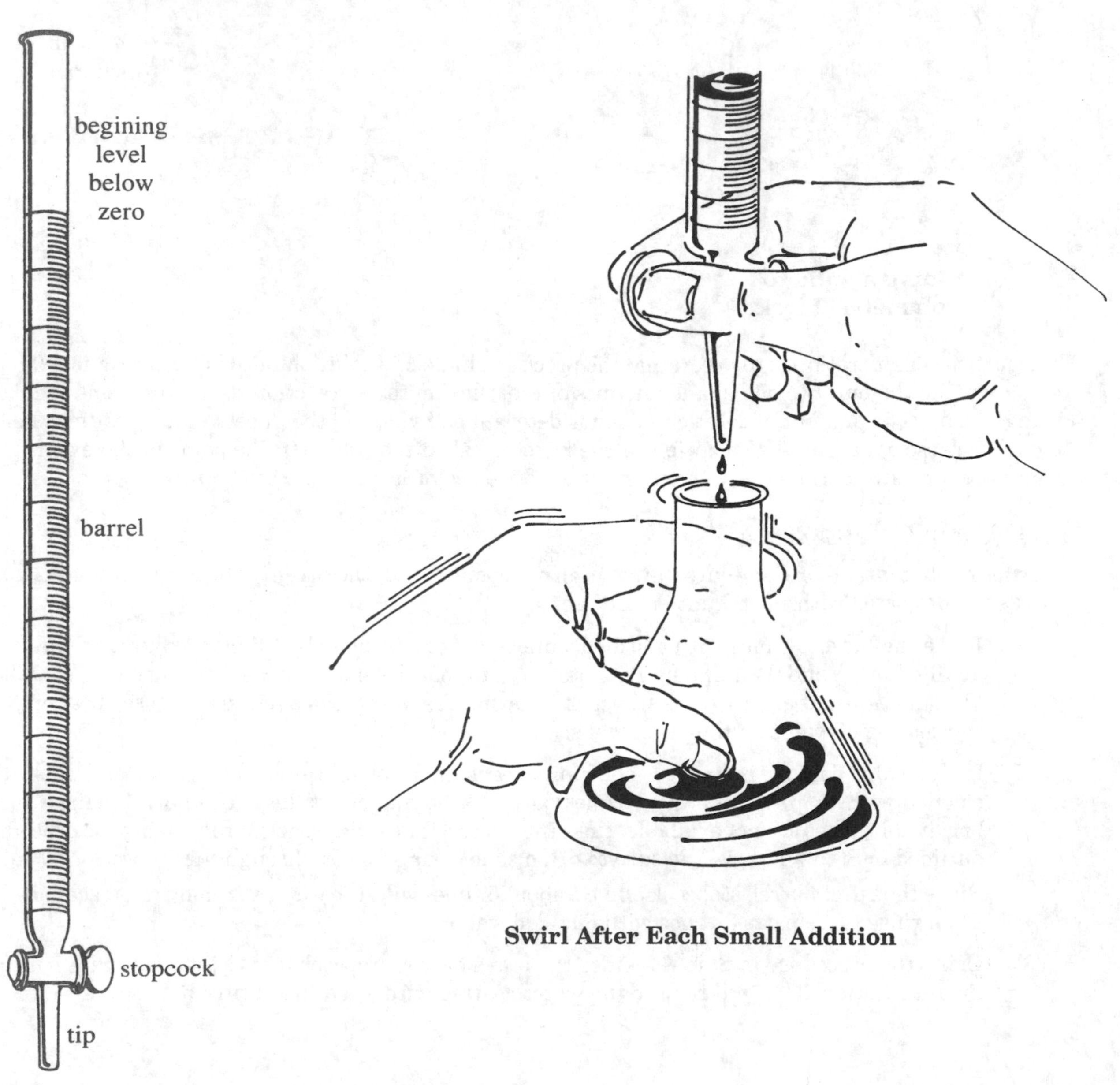

Swirl After Each Small Addition

POINTERS ON TITRATING

Following are some "tricks of the trade" that can improve the accuracy of your work.

8. Make your buret readings as precisely as possible. Determine the readings to a hundredth of a milliliter by estimating the distance between volume markings. Wait 15 sec. before making a reading. This allows the solution on the sides of the barrel to finish draining. If you find it difficult to see the exact bottom of the meniscus, place a darkened piece of paper (see figure) behind the buret and about 1 cm below the meniscus. The meniscus will "reflect" this darkening and become more distinct.

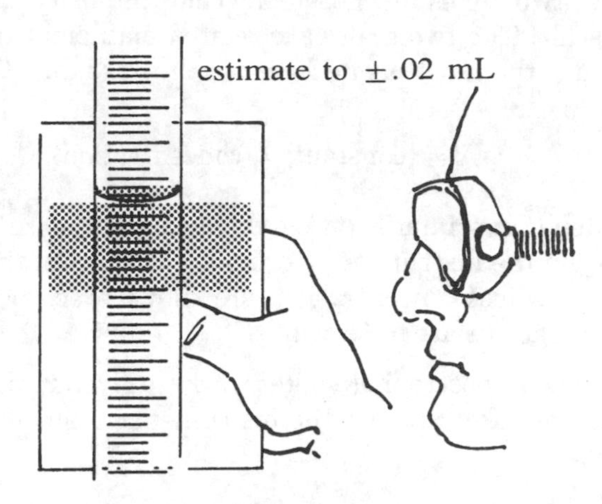

Making the Meniscus Clearer

9. One should locate the end-point within 1 drop (1/20 mL). Adding one drop at a time through the entire titration takes a great deal of time and is boring. There is a way around the problem. Begin by adding about ½ to 1 mL at a time and watch the coloring where the titrant hits the solution. If the color change dispels by itself, the end-point is not close. If you must swirl the flask to dispel the color change, you are getting closer and must cut down the size of the addition. When 2 or more swirls are needed to dispel the color change, you are very close and should make single-drop additions. Alternatively, if you have a reasonable idea of where the end-point should be, add 80% of the needed titrant in one shot and then begin your dropping additions.
10. Place a white sheet of paper under the flask. This will allow you to see the color changes more clearly.
11. Near the end of the titration, rinse down the inside of the flask with a little distilled water to return all chemicals to the body of the solution.
12. Do not trust the results of a single titration. If possible, repeat the titration of a particular solution a couple of times to check on the reproducibility of your result.

D.2 Spectrophotometer Measurements

A spectrophotometer is an instrument that measures the color intensity of a solution. The color intensity can then be related to the concentration of the solution. The basic principles involved are used by each of us in our everyday lives. For example, we judge the strength of tea or coffee by its color. The darker the color, the stronger the drink. We can rephrase this to say: The more intense the color, the higher is the solution concentration. The spectrophotometer does just what your own eyes do: it acts as a color intensity meter. The advantage of the instrument is that its measurements are much more precise than those of your eye.

Most spectrophotometers have two scales for measuring color intensity: an absorbance scale (A) and a percent transmittance (%T) scale. The two scales are related mathematically as shown later. However the absorbance (A) readings are the most useful because they are directly proportional to the molar concentration of the solution.

$$A = (\text{constant}) \times (\text{concentration})$$

The constant in the equation depends primarily on the chemicals used in the solution (solute and solvent) and on the wavelength setting on the instrument. Normally the value of the constant is calculated from absorbance readings of solutions whose concentrations are known precisely. This value is then used with similar solutions whose concentrations are unknown.

A more detailed discussion of how a spectrophotometer works is given in part B. It is suggested that you study this section both before and after you use the instrument in lab. This will help put your data in perspective.

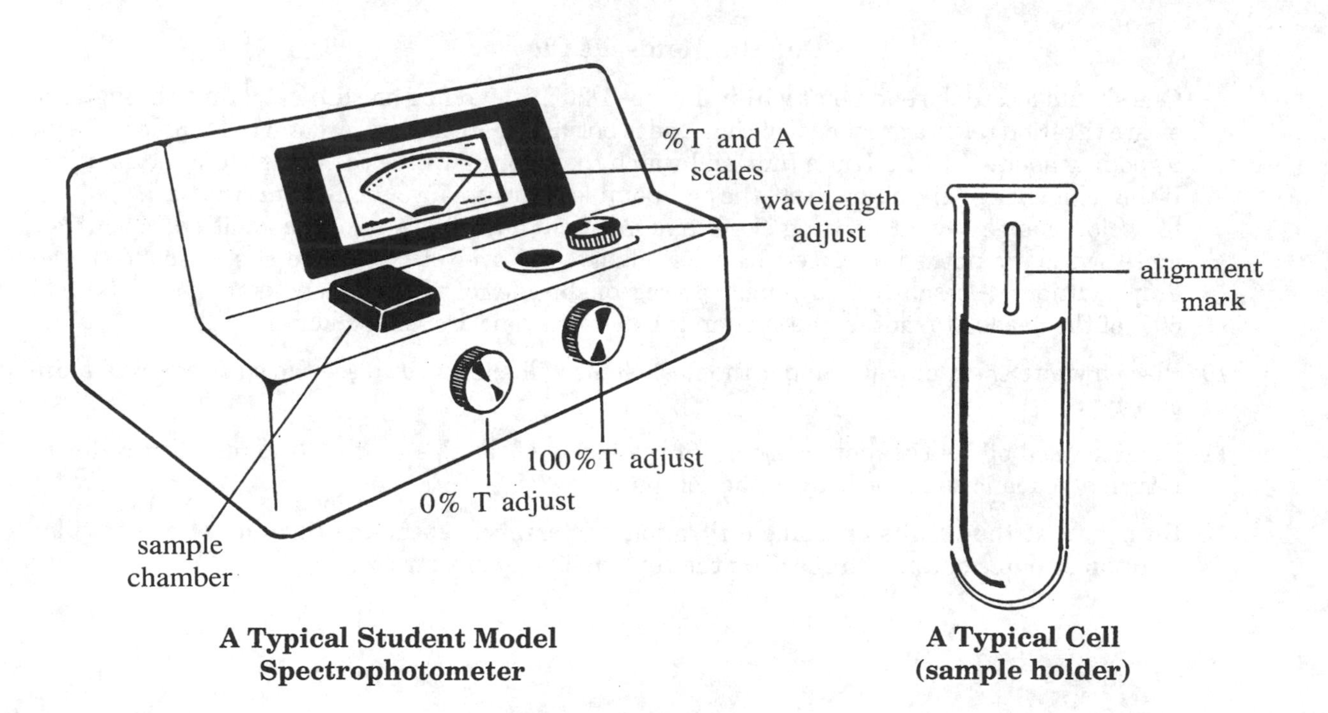

A Typical Student Model Spectrophotometer

A Typical Cell (sample holder)

A. PROCEDURE

These are general procedures for making precise measurements. Study them and incorporate any changes given by your instructor for your specific spectrophotometer.

There are three things the experimenter must watch if precise measurements are to be obtained: (1) that the instrument is standardized; i.e., gives proper readings at each end of the scale, (2) that the solutions measured are not contaminated or diluted, and (3) that the outside of the cell is spotlessly clean and dry.

1. Turn on the instrument and adjust the wavelength knob to the required setting. Allow the instrument to warm up for the required amount of time.
2. Rinse the cell 4 to 6 times with distilled water. Fill it about 3/4 full with water. Holding the cell at its top, wipe the entire outside dry with tissue or a clean towel. Check the outside for smudges and liquid. If any exist, remove them. From now on handle the cell only at its top.
3. With no cell in the instrument set the reading at exactly 0%T ($A = \infty$) using the proper knob.
4. Insert the water cell into the cell compartment so that the mark on the cell is properly aligned. Close the compartment lid. Using the proper knob adjust the reading to exactly 100%T ($A = 0$).
5. Repeat steps 3 and 4 until proper readings are obtained at both 0%T and 100%T.
6. Rinse the inside of the cell at least twice with small portions of the solution to be measured. (The purpose of this is to prevent dilution by coating the entire inside of the cell with the solution.) Fill the cell about 3/4 full with the solution. Holding the cell at its top, rinse the lower 3/4 of its outside with distilled water. Wipe the entire outside of the cell, removing all liquid films and smudges.
7. Holding the cell at its top, insert it into the cell compartment. Align it to the mark and close the compartment lid. Once the reading stabilizes, record it as directed (A or %T).

Other solutions can now be measured by repeating steps 6 and 7. It is wise to check the standardization of the instrument (steps 2 to 5) about every 20 minutes. If you have enough solution, it is also recommended that you make 2 or 3 separate measurements of each solution to obtain an average reading.

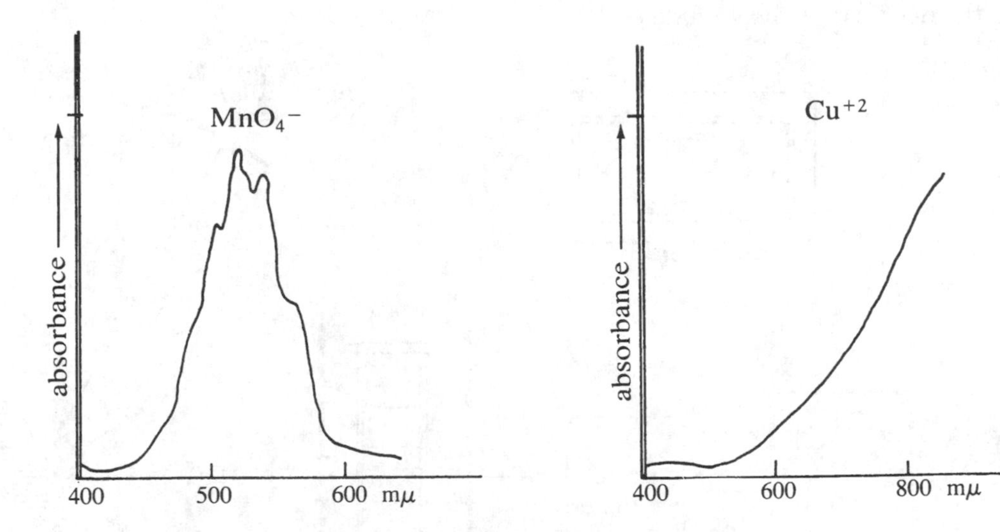

Absorbance Versus Wavelength

B. DETAILS OF LIGHT ABSORPTION

The reason why some solutions are colored is that selected portions of the light passing through them are absorbed (captured) by the solute molecules. Several factors affect the portion of light and the fraction of light absorbed.

WAVELENGTH

Light, also called electromagnetic radiation, is pure energy. It may be viewed as packets of energy, called photons, traveling with a wave motion. The energy (E) of each photon gives its motion a characteristic wavelength (λ). The relationship is $E = hc/\lambda$ where h is Plank's constant (6.62×10^{-34} joule - sec.) and c is the speed of light (3.00×10^8 m/sec). Thus light of longer wavelength has photons of lower energy.

A molecule will absorb light whose energy corresponds to one of its own energy needs. Thus some wavelengths of light are heavily absorbed while others remain untouched. The variation in absorbance with λ is illustrated for two ions on page 325. The visible spectrum is light having λ between 400 and 800 nanometers (1nm = 10^{-9} m). Our eyes "see" these different wavelengths as different colors. (Light having all these wavelengths is clear or "white" light.) When looking at a solution we see only those colors (wavelengths) not absorbed by the solute molecules. Thus a Cu^{2+} solution is blue because yellow, orange and red wavelengths have been absorbed.

NATURE OF THE CHEMICALS

Visible light has energy matching that needed to move electrons to higher energy levels in many molecules. However, since the electron energy levels in each type of molecule have slightly different spacings, each chemical species will show a different pattern of absorbance versus λ.

CONCENTRATION OF THE SOLUTION

As the concentration rises, the light photon has a greater chance of encountering an absorbing molecule as it passes through the solution. Thus the fraction of light absorbed increases with higher concentrations. The absorbance also increases with the ***length of the cell*** for much the same reason. If the light photon must pass through more solution, the probability that it will be absorbed increases.

In a light-measuring experiment, the experimenter must have a method of controlling these factors if the data are to be useful. The spectrophotometer is designed to give such control. A simple schematic of a spectrophotometer is below. White light from a high intensity lamp is focused on a grating or prism, which fans out the light according to its various wavelengths. The grating can be rotated so that light of just the required wavelength hits the sample. The light passes through the sample and hits a light detector, called a photomultiplier. The detector converts the light intensity into an electrical current, which moves the needle over the reading scale.

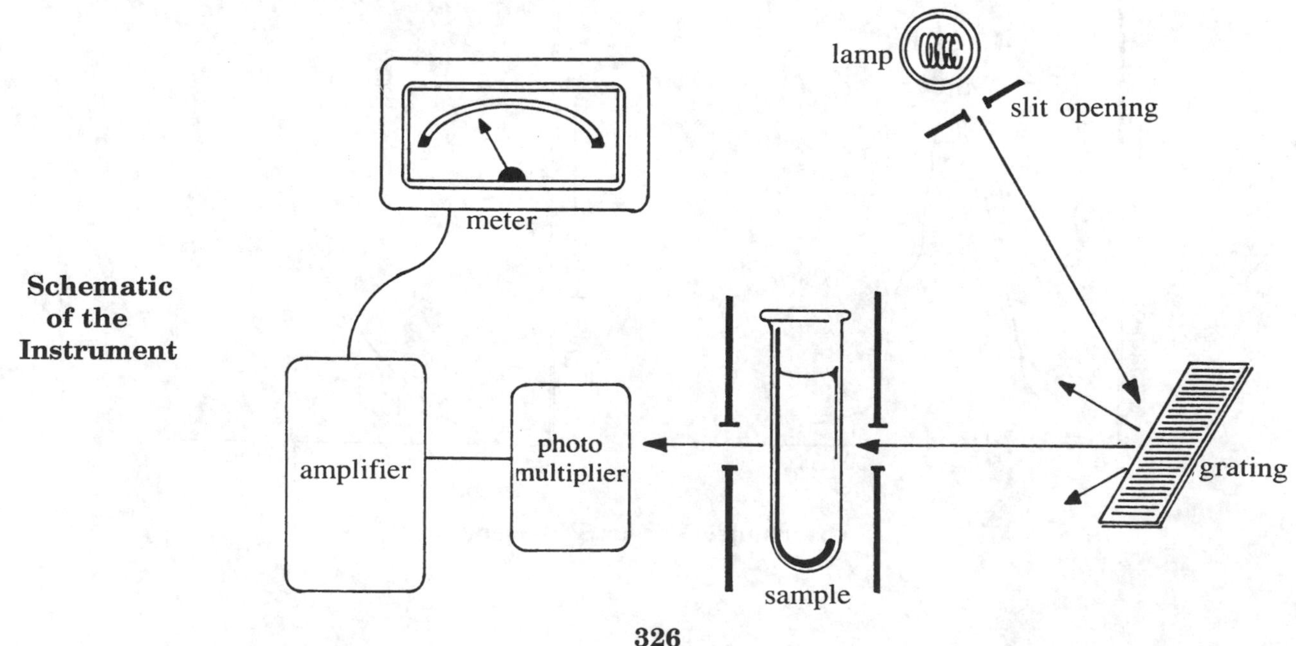

Schematic of the Instrument

In standardizing the instrument, the needle is set to O%T (A = ∞) with no light reaching the detector, i.e., intensity I = 0. It is set to 100%T (A = 0) when the detector senses I_{max}, the light intensity emerging from the cell and water. This is the maximum amount of light that can emerge from any sample. If the solution sample is now inserted, the scale will read a light intensity, I, that is properly compared to the no light and to the maximum intensity situations.

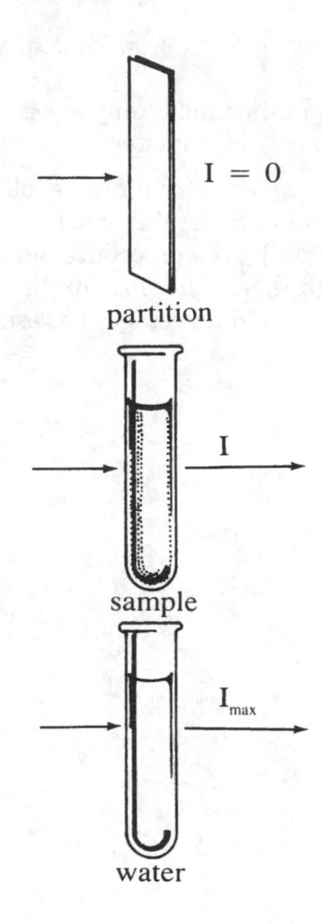

The Two Standardization and the Sample Readings

The mathematical relationship for light absorption is as follows:

$$\%T = \frac{I}{I_{max}} \times 100$$

Thus the percent transmittance measures the ratio of the intensity emerging from the solution to the intensity emerging from water alone. It is an indication of the fraction of light captured by the solute. The higher the fraction of light captured, the lower the %T.

Unfortunately, these intensity fractions do not relate simply to concentrations. They are related through the common log function (i.e., see pHs).

$$-\log \frac{I}{I_{max}} = E \times d \times c$$

c is molar concentration of the solution, d is the length of the light path through the solution (i.e., the inner diameter of the cell), and E is a constant for the particular solute and wavelength used. E is called the extinction coefficient.

Because finding logs of numbers is tedious, the reading scale of most spectrophotometers are marked to give the log function directly. This scale reading is called Absorbance, A.

$$A = -\log \frac{I}{I_{max}}$$

$$A = E \times d \times c$$

$$A = (\text{constant}) \times c$$

It must be remembered that this last equation holds only when E and d are constant—that is, where the absorbing compound, the wavelength and the diameter of the cell are not changed.

The numerical value of the "constant" can be determined as follows: Make solutions of precisely known concentrations of the compound in question. Several should be made, each with a different concentration. Measure the absorbance of each solution by the procedure outlined above. Plot the data as A versus c and draw the best straight line through them. The slope of the line is the "constant." Alternatively you can determine A/c for each known solution and average the results to get the best value.

D.3 pH and Conductivity

A. pH

The Concentration of H^+ in moles/liter in water solution, $[H^+]$, is controlled by the following reaction:

$$H_2O \rightleftarrows H^+(aq) + OH^-(aq).$$

This reaction does not proceed to completion, and the concentrations of $[H^+]$ and $[OH^-]$ are very small. Because $[H^+]$ generally have values between 10^{-1} and 10^{-14}, it is convenient and more compact to express these numbers as logarithmic functions of the H^+ concentration. Since these numbers are less than one, their logarithmic function will be negative, thus it is also convenient to change the sign. This compact method of expressing H^+ concentration is called pH and is defined by the following equation:

$$pH = -\log[H^+].$$

In order to determine the pH of a solution, determine the concentration of H^+ in moles/liter, take the logarithm and multiply the number (usually negative) by -1.

In order to determine the $[H^+]$ of a solution given the pH, multiply the pH by -1 and take the antilog of the number: $[H^+] = 10^{-pH}$.

The pH of a solution can be approximately determined using pH paper or indicators, or more exactly by using a pH meter.

B. THE pH METER

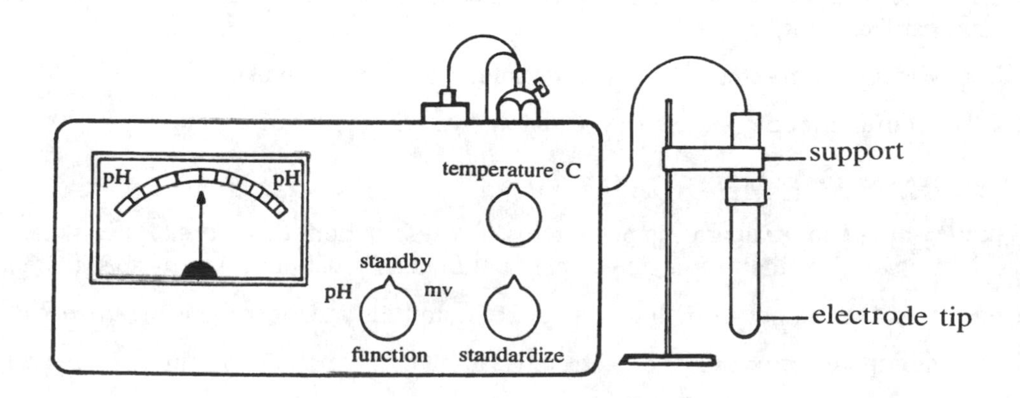

Schematic of the Instrument

A pH meter consists of two electrodes that generate an electrical signal that can be read on a meter in pH units. One of the electrodes is sensitive directly to $[H^+]$. It is called a glass electrode, and it has a very thin, spherically shaped glass membrane at its tip. The other is a reference electrode, which completes the electrical circuit. Reference electrodes will clog up if allowed to dry out. If this occurs, see your lab instructor about what to do. A complete circuit depends on both electrodes having their tips completely submerged in the solution to be measured.

Your pH meter may be equipped with a combination electrode, which has both electrode systems in a single probe. These two electrodes are still connected to each other externally through the glass electrode tip and a small pin hole or wick on the side of the tube. Make sure the probe is covered with solution to depth sufficient to cover the pin hole.

MAINTENANCE AND CARE OF pH METERS AND ELECTRODES

The electrode tip is the most delicate part of this instrument. Be extremely careful that nothing touches it except liquids and tissue for drying. Do not allow it to touch the bottom of a beaker. When drying it with tissue, ***do not rub dry—pat it dry gently.***

Each pH meter should come equipped with the following auxiliary items: one box of tissue, one squeeze bottle of distilled water, one container of buffer solution, one container of soaking solution, and one large beaker to collect waste wash solution.

At the beginning of each lab, check the meter for these items:

1. The power plug should be firmly seated and grounded in the socket. If the meter needle sways wildly (in the pH mode) when you pass your hand over the meter, the ground is not secure.
2. The electrode input jacks should be firmly seated all the way in their sockets.
3. Check the electrodes for cracks. Cracked or severely scratched electrodes do not work.
4. The electrodes use liquid junction references. Therefore, the reference electrodes and the outer jackets of combination electrodes must be kept full with saturated KCl. If the level is low, fill through the hole in the body below the cap until about ¼ inch below the hole. The holes should remain open. Some electrodes do not require filling.
5. Buffer solutions should be checked with pH paper to see if they are still reliable. If there is any doubt, get fresh buffer.
6. Electrodes should be stored in a soaking solution when not in use.
7. Report nonfunctioning meters to your lab instructor.

INSTRUCTIONS FOR MEASURING pH

Many pH meters have similar operating procedures. The instructions below are for pH meters as pictured on the previous page. Your insturctor may supply different instructions for your specific meter.

1. Check that the temperature dial is set at 20 °C and the function dial is on ***standby.***
2. Move the support up so that the electrode tip is out of the soaking solution. Set this solution aside.
3. Using the squeeze bottle, rinse the bottom third of the electrode, catching the water in the waste beaker. Gently ***pat*** away the excess water from the electrode, using a clean tissue.
4. Lower the electrode into a beaker of buffer solution. Only one inch of the electrode need be immersed. Be careful to not bump the electrode tip on the sides or bottom of the beaker.
5. Move the function switch to pH. Turn the ***standardize*** dial until the meter reads the pH of the buffer solution.
6. Move the function switch to ***standby.***
7. Repeat steps 2 and 3 to again rinse the electrode. Set aside the buffer solution and be sure not to contaminate it.

8. Lower the electrode into the sample to be measured. Turn the function switch to pH and record the reading. Return the function switch to ***standby.***
9. Repeat steps 2, 3, and 8 to obtain reading of other samples.
10. When readings are complete, rinse the electrode as in steps 2 and 3. Immerse the electrode in the soaking solution (0.01 M HCl). Be sure the function switch is on ***standby.***

C. THE CONDUCTANCE METER

Conductance measures the ease with which electrical current can pass through a solution. By definition, conductance is the reciprocal of the electrical resistance of the solution (common units are mhos or μS, micro-Siemans).

Pure water does not conduct electricity; only solutions with ions dissolved in them will conduct. The higher the charge on the ions dissolved and the higher their concentration, the higher will be the conductance of the solution. Thus conductance measurements can be used, qualitatively, to tell the difference between types of ions (for example, +1 versus +2 cations) or to tell the difference between two concentrations of the same salt.

Several kinds of conductance meters are available; two of the easier ones to use are shown below. Your instructor will give specific directions for using your meter, but they will generally follow this order:

1. Shut the meter off. ***Disconnect the power cord on the light bulb apparatus.***
2. Rinse the electrodes with distilled water, and pat them dry.
3. Immerse the electrodes in the solution to be measured.
4. Turn the meter on. Wait a few seconds for the reading to stabilize, and record the reading.
5. Turn the meter off (***disconnect that power cord***) before pulling the electrodes from the solution. Repeat steps 2 to 4 for other solutions.

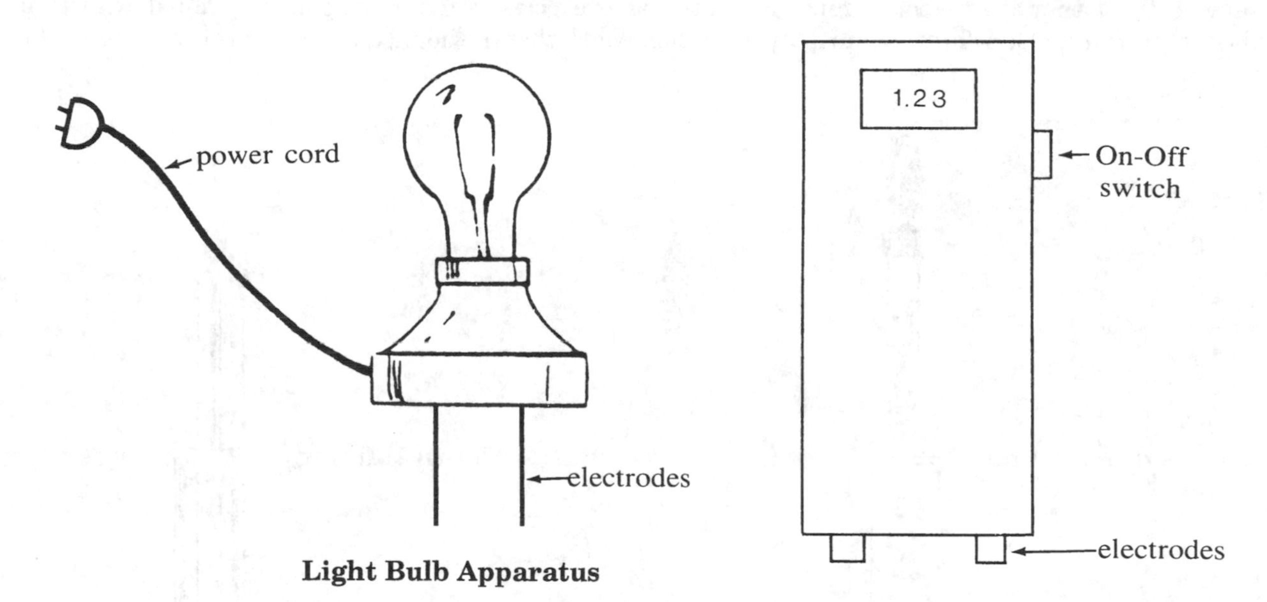

Light Bulb Apparatus

Digital Conductance Meter

D.4 Qualitative Ion Analysis Using Semimicro Techniques

In these experiments you will be concerned with determining what particular ions are in a solution or solid. This concern with only what is present is called ***qualitative analysis***. (A concern for how much is present is called ***quantitative analysis***). You will be working with very small amounts of solutions, usually about ¼ to ½ of a milliliter. Thus the experimental procedures are called semimicro (small amount) techniques. Two other definitions of importance are:

precipitate—a solid formed when two solutions are mixed (abbreviated ppt.)

supernatant—the liquid phase of a solid plus liquid mixture (abbreviated spn.)

The basic idea in these experiments is to add a reagent to a solution of ions and to note any and all changes that occur. In particular one looks for and notes down

1. the formation and color of a ppt.
2. changes in color of the solution or spn.
3. redissolving of a ppt.
4. evolution of a gas.

A. ADDITIONS AND MIXING

The amounts of solutions to be mixed in each test are often critical to the results and must be carefully watched. The amounts are usually controlled by counting drops of solution used (18 drops = 1 mL). When making additions, suspend the end of the dropper just above the opening of the test tube and gently apply pressure to the dropper bulb so that the drops fall one at a time into the test tube.

When test reagent has been added to an ion solution, mix the two completely before you note down your observations. The results that you see when one solution is layered on top of the other may not be the same as when they are mixed. To mix the solutions move a clean stirring rod gently up and down within the test tube; i.e., use a "butter churn" type motion. Watch that the liquid does not overflow the test tube.

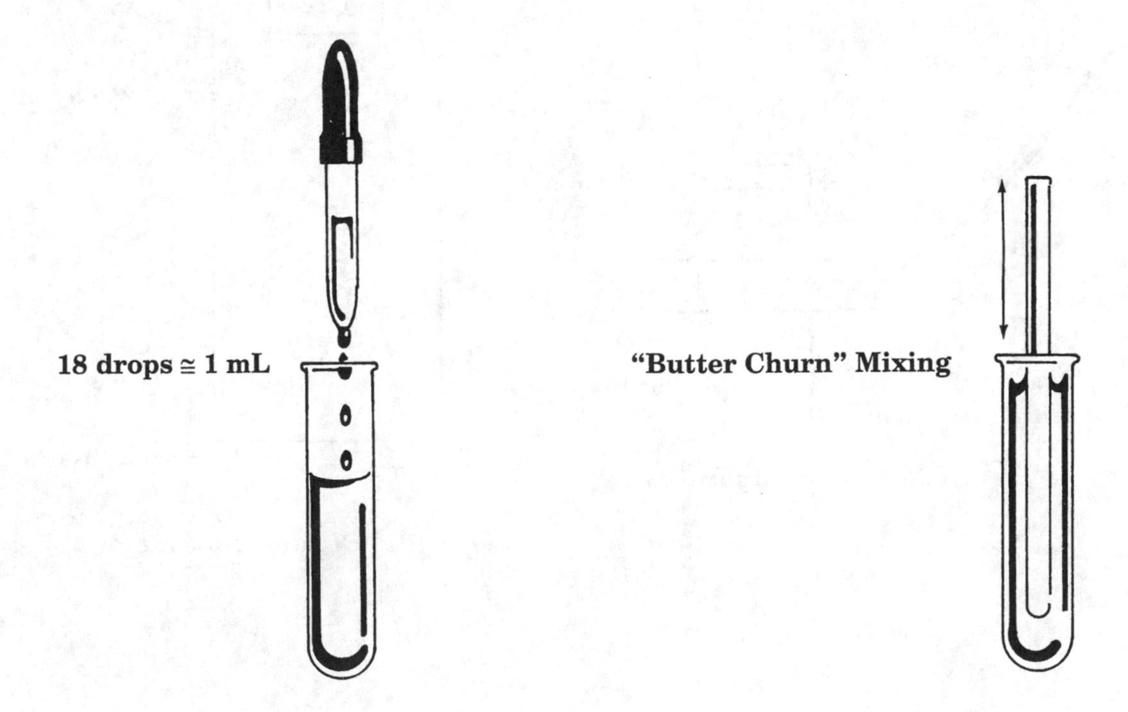

B. THE CENTRIFUGE

The centrifuge is a machine that spins a test tube solution at high speeds, causing the heavier and lighter parts of the mixture to separate. In essence, it accelerates what gravity would have done anyway. The centrifuge is used to separate a precipitate from the supernatant. Usually only one or two minutes of centrifuging are needed to separate the ppt. and spn.

Precautions to be observed in using the centrifuge are as follows: (1) Do not have the test tubes more than 4/5 full. This will prevent spilling in the centrifuge. (2) Place the tubes in the centrifuge in a symmetrical pattern, such as one tube directly across from another. If you are centrifuging just one test tube, place another containing water across from it. If the placement of the tubes is not balanced, the spinning will be uneven and the centrifuge will move ("walk") on the bench. (3) Allow the centrifuge to slow down by itself. Helping it slow down can cause injury. Also if the slow-down is too abrupt, the ppt. and spn. will remix.

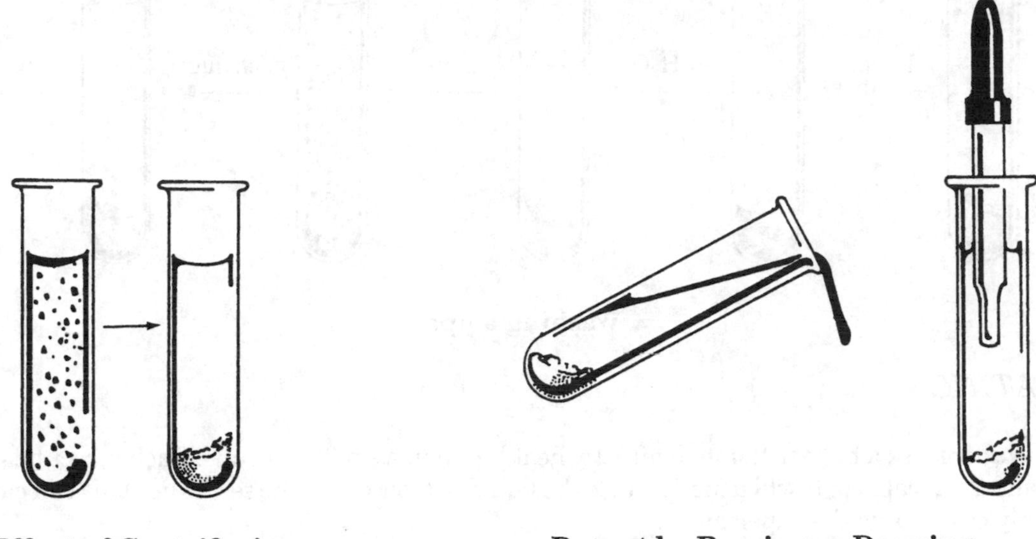

Effect of Centrifuging **Decant by Pouring or Drawing**

C. DECANTING

This is the term used to denote removing a supernatant from a precipitate. There are two decanting methods. The first is to simply pour the spn. out of the test tube, leaving the ppt. behind. The second method is to draw the spn. out of the test tube, using a dropper. This dropper method is used when the ppt. is not a firm pack and might pour out with the spn. It is also the preferred decantation method where the spn. is to be saved for later use. It is often difficult to pour directly from one small test tube to another.

When using the dropper method, be sure to compress the bulb ***before*** inserting the dropper. A rush of bubbles in the spn. will cause the ppt. and spn. to remix.

D. WASHING A PRECIPITATE

Not all of the supernatant is removed from the test tube when one decants. A few drops remain on top of the precipitate, some clings to the test tube walls, and some is trapped within the packed precipitate. If a later test on the ppt. requires that the chemicals in the spn. be absent, these traces of spn. are removed by a procedure called washing.

Washing a ppt. involves the following sequence of steps: (1) Decant the original spn. (2) Add to the ppt. about 1 mL of distilled water (or other prescribed liquid). (3) Completely mix the H_2O and ppt. This allows the H_2O to leach out any spn. trapped in the packed ppt. (4) Centrifuge the mixture. (5) Decant the liquid, discarding it. (6) Repeat the process. Usually two such water washes are sufficient as they will leach out about 97% of the spn. originally left in the test tube.

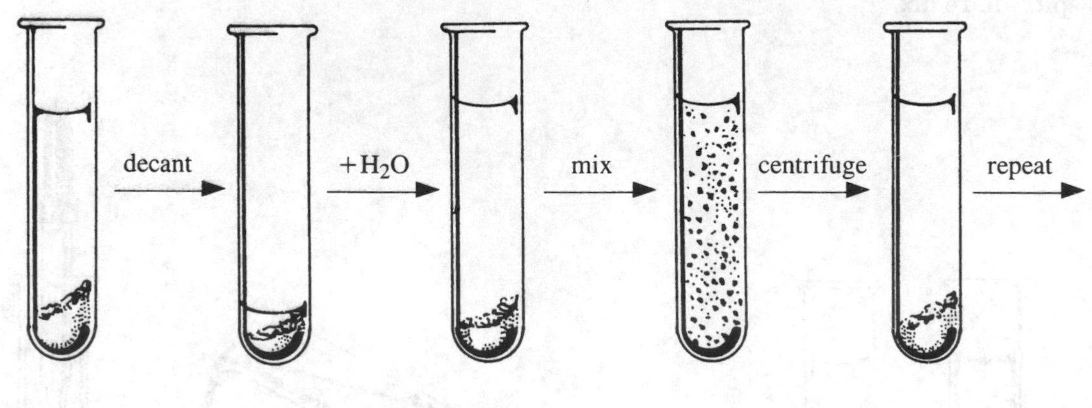

Washing a ppt.

E. HEATING

You ***should not*** use a burner flame to directly heat liquid in a small test tube. Such direct heating is too intense and too localized. It will usually cause the liquid to blow out of the test tube. This can cause injury and, in any case, ruins the experiment.

Heating is to be accomplished by resting the test tubes in beakers of heated water. This heating is more even and keeps the temperature from exceeding 100 °C. If a test tube heating platform (also called an aluminum holder) is available, it should be used in the beaker. Otherwise use a beaker narrow enough to keep the test tubes protruding over the beaker lip.

Adjust the water level in the beaker so that only the lower 1/2 to 3/4 of the test tube is submerged. Move the burner flame in and out as needed to keep the water hot but to prevent violent boiling.

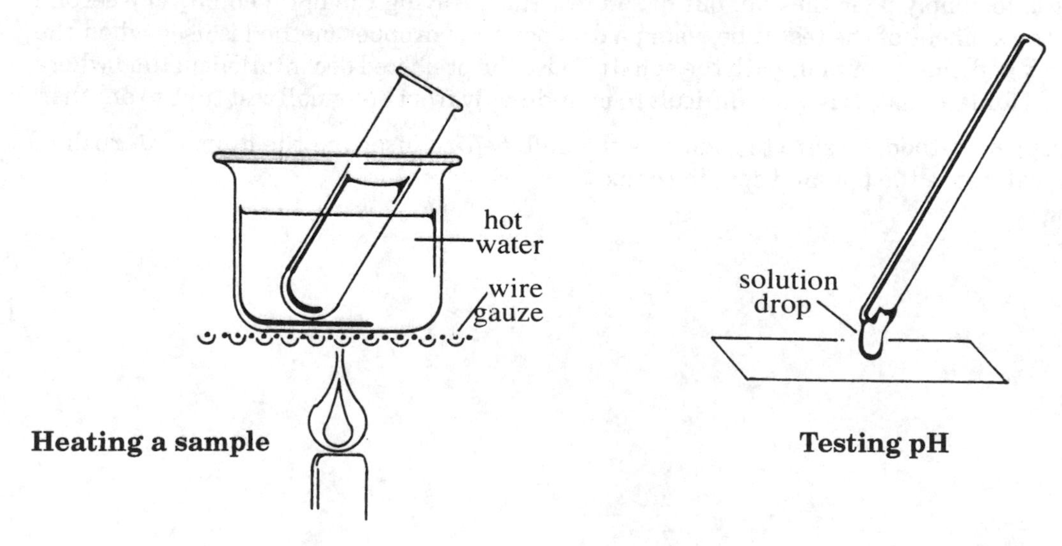

Heating a sample **Testing pH**

F. pH MEASUREMENT AND CONTROL

It is important in many reactions that the solution's pH be carefully controlled. (See Appendix D.3 for an explanation of pH.) The control is achieved by adding an acid or base solution dropwise until the required pH is obtained. The pH of the solution should be measured after each dropwise addition by using litmus or pH paper. Do not dip the pH paper into the solution. This can lead to chemicals from the paper getting into the solution. The proper procedure is to mix the solution with a stirring rod and then to touch the wet end of the rod to the test paper.

The pH of a gas bubbling out of a solution can be tested by wetting the test paper with water and then holding it over the mouth of the test tube.

G. CONTAMINATION

Even the smallest contamination of your ion or reagent solutions can change the test results. Contamination is a particular problem with these experiments because you reuse equipment so often. ***Clean and distilled-water rinse each piece of equipment after each use.*** Thus you should clean a dropper before using it in a different solution, clean a stirring rod before using it to mix a second solution, etc.

Most often cleaning requires only that the test tube, rod, etc. be rinsed several times with tap water, then 2 to 3 times with distilled H_2O. The equipment does not have to be dry. Use a test tube brush to clean out stubborn ppts. A dropper can be cleaned by removing the bulb and running water down the barrel.

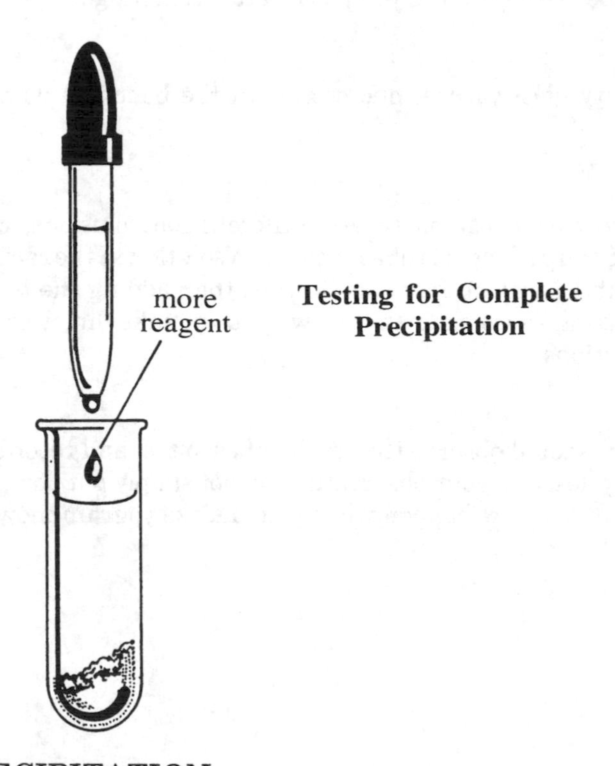

Testing for Complete Precipitation

H. COMPLETE PRECIPITATION

In Section Two of each "L" experiment you will be analyzing a solution containing more than one ion. (You will create an analysis scheme—a game plan—for this work.) In this analysis of an unknown you will often depend on precipitating an ion completely from solution to separate it from the other ions. It is important that this precipitation reaction be complete.

After centrifuging the ppt. plus spn. mixture, add a drop or two more of the test reagent. If new ppt. forms in the spn., add two more drops of test reagent, mix and recentrifuge. Again add a drop of reagent. Continue this 1- to 2-drop testing of the spn. until no new ppt. forms. At this point the reaction is complete.

I. ORGANIZATION AND REAGENTS

It is strongly suggested that you take a few minutes to plan your work, especially when doing the Section One experiments. You will be performing 25 or more individual tests in one period; thus, some organization is needed to save time and to not confuse your results. The following ideas might be helpful.

1. When taking chemicals from the stock bottles, immediately label the beaker used with both the reagent solution's ***name*** and ***concentration***. Often you will have to use the same chemical at different concentrations at different times. Take only the small amounts of stock reagents needed (usually about 5 mL).

2. After using a dropper, immediately return it to its reagent beaker. Do not have two droppers out of their reagent beakers at one time. This will keep you from returning a dropper to the wrong reagent beaker.

3. Label each test tube and assign it a labeled position in your test tube rack. Never have two test tubes out of the rack at the same time; always return one before picking up another. This keeping track of test tubes is needed because the label on the tube can be smeared in handling. Be sure the label on a tube is clear before you put it into a centrifuge.

4. Put all used, dirty glassware in one section of the bench so as not to confuse it with clean glassware.

5. When you are to run a test on several different ion solutions, run the test on all of them simultaneously. Set up a labeled tube for each ion solution in the rack. Put the proper ion solution into each test tube. Next, go from one to the other, adding the test reagent and mixing. This simultaneous testing saves time and allows you to make direct comparisons of the results for different ion solutions.

6. Both lab partners should observe the results of each test and record their own perception of the results. Also fully describe your observation, do not simply put down "Reacted." You will need to recall exactly what you saw happen when you analyze your unknown.

SAMPLE ANALYSIS SCHEME OR FLOW CHART

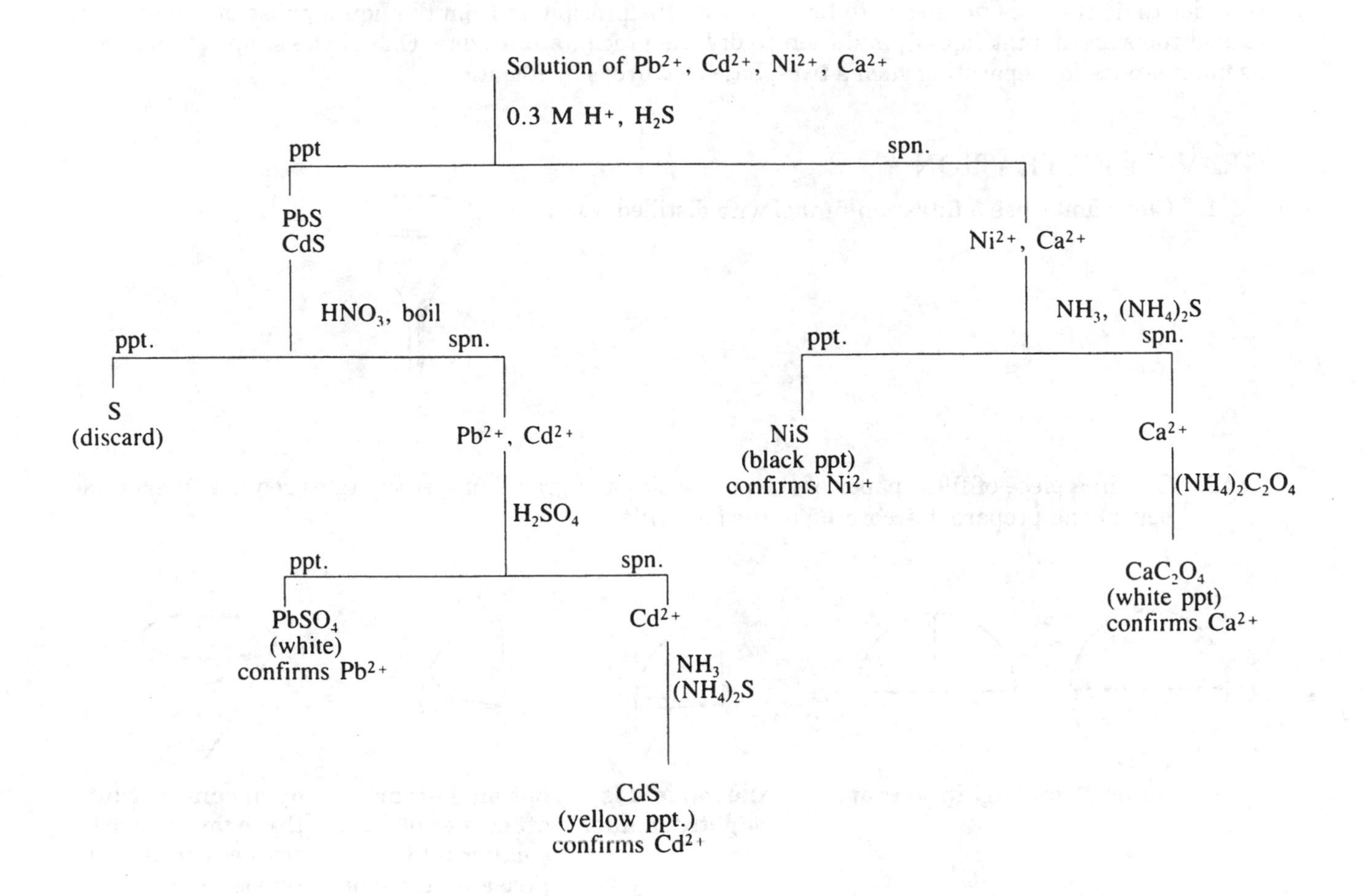

D.5 Filtrations

A large number of chemical reactions in solution result in the formation of an insoluble solid. It is sometimes necessary to weigh the quantity of solid (called a precipitate) to determine the amount formed. In order to do this it is necessary to first separate the precipitate from the liquid phase of the mixture (called the supernatant liquid), and then to dry the precipitate residue. One of the simplest and most useful methods for separating such a two-phase mixture is by filtration.

GRAVITY FILTRATION

1. Clean and rinse a filtration funnel with distilled water.

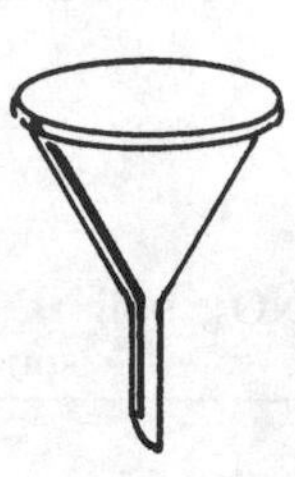

2. Obtain a piece of filter paper of the proper size and grade (fine precipitates require finer-grade paper) and prepare it according to the following:

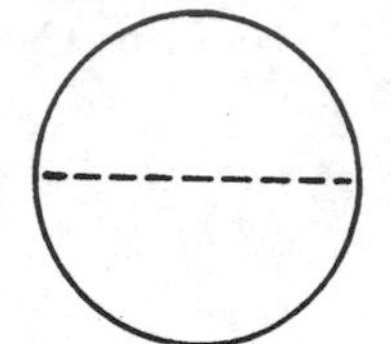

fold in half

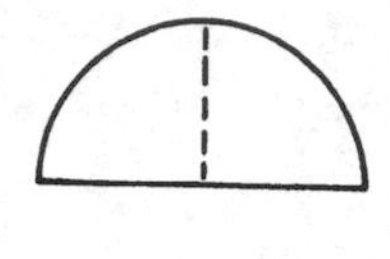

fold again so that . . .

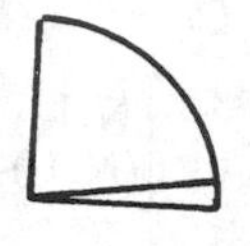

the top fold is slightly smaller

optional—tear off corner of smaller fold (to give a better seal)

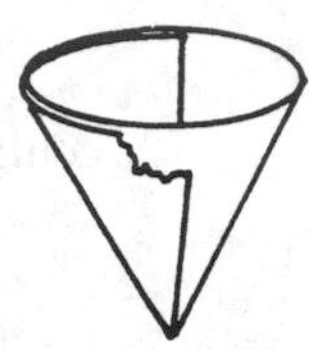

open cone so that three layers (with torn cover) are on one side

3. If the filter paper needs to be weighed, do so after the corner is torn off and before wetting it.
4. Place the cone in the damp filter funnel and wet with distilled water. Gently press the upper edge of the filter paper against the funnel to form a tight seal. The funnel stem should hold a column of water without bubbles. This condition allows maximum speed in filtration.
5. Gently pour the solution to be separated into the filter, being careful not to overflow the top of the filter paper.

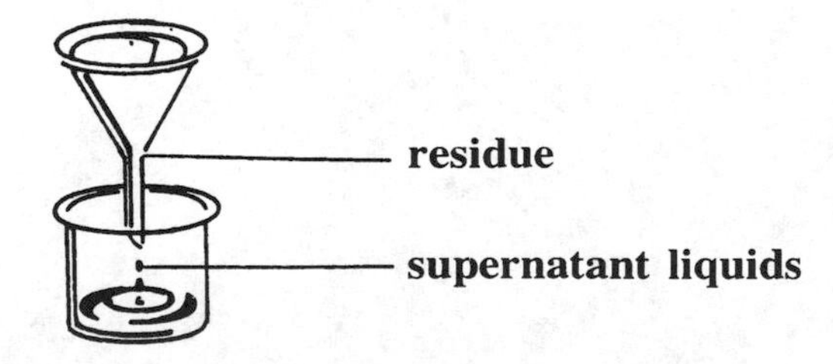

6. Wash the residue with distilled water. If there is a danger of losing some of a partially soluble residue wash with small amounts of the supernatant liquid or another liquid which will not dissolve the residue.

7. Sometimes a precipitate may be too fine to be caught in the filter paper. If the precipitate passes through use one of the following procedures.

   a. Filter the supernatant with passed precipitate again. Often the residue in the filter paper will act to filter the passed precipitate during a second or third filtration.

   b. Use a finer grade filter paper.

   c. Use a double thickness of filter paper. Don't forget to add the weight of the second piece of paper if your experiment requires weighing.

   d. Digest the precipitate by boiling it for a period of time (20 minutes) before filtering. This increases the size of the precipitate crystals, making them easier to filter. Before carrying out this procedure, check with your instructor to make sure heating your precipitate will not destroy your experimental results.

SUCTION FILTRATION

If the gravity filtration of a precipitate is particularly slow, suction filtration using a water tap aspirator can be tried. Consult with your instructor for approval and advise. A possible equipment set up is shown below.

Two disadvantages of aspirator filtrations are that they are cumbersome to use, and the supernatant might get contaminated with tap water. Some precautions to observe are listed below.

1. The filter paper should lie flat in the Buchner funnel and must cover all the holes.
2. The aspirator should be turned on ***before*** it is connected to the filter flask.
3. When the filtration is complete, the tubing should be disconnected from the filter flask ***before*** the aspirator is shut off.

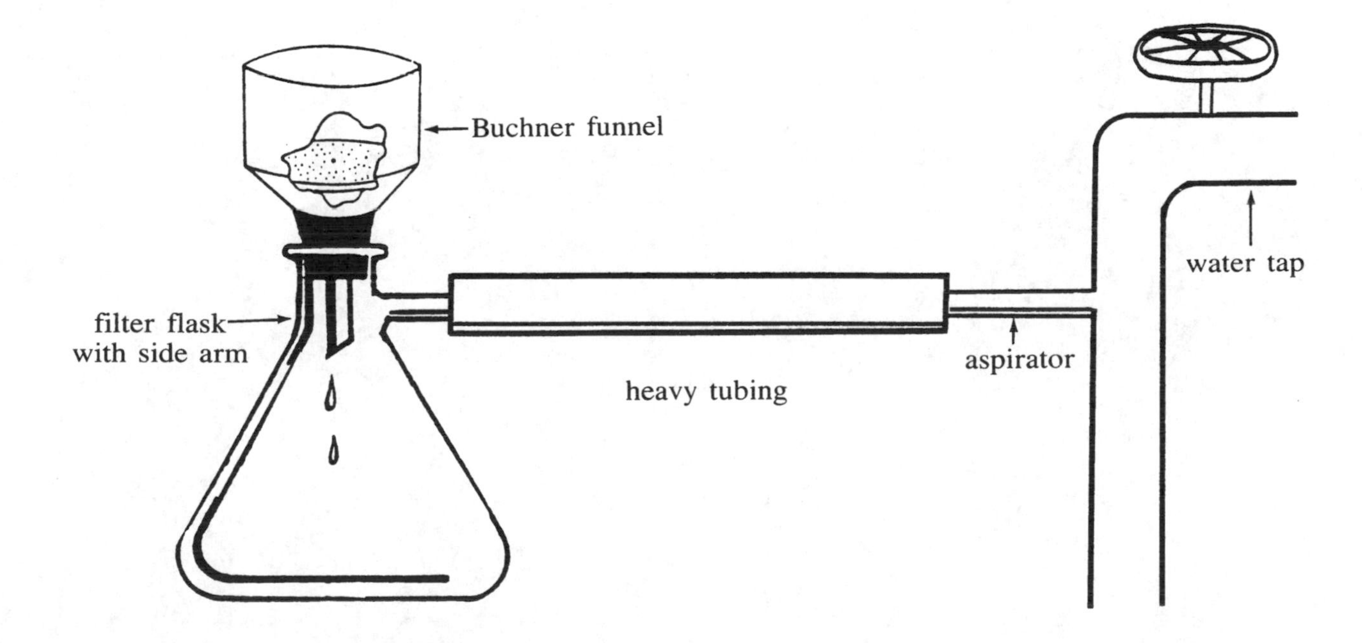

DRYING THE RESIDUE

Once the mixture is filtered and washed there are several methods that can be used to dry the residue. The residue can sometimes be rinsed with a volatile solvent like acetone that will wash the water away and evaporate faster than water. Care must be taken, however, that the residue is not dissolved or washed away with the solvent. The filter paper should be carefully removed from the funnel and unfolded and placed on a paper towel or other absorbent material. The residue should be spread on the paper, being careful not to lose any. The residue can be dried in a number of ways.

1. ***Air-drying.*** Allow the residue and paper to stand in a well ventilated space (if possible) for a period of time (overnight).
2. ***Hot water bath.*** If the residue will not decompose at 100 °C, place it and the filter paper in a watch glass. Place the watch glass on a large beaker half filled with water. Boil the water for 10–15 minutes. Remove the watch glass and allow to cool.
3. ***Drying oven.*** If the residue will not decompose at the temperature of the drying oven, place it and the filter paper on a watch glass and place in the oven for 20–30 minutes.

The simplest procedure for checking if the drying process is complete is to weigh the sample, dry it again, and then reweigh the sample. If the second drying causes no weight change, the sample is completely dry.

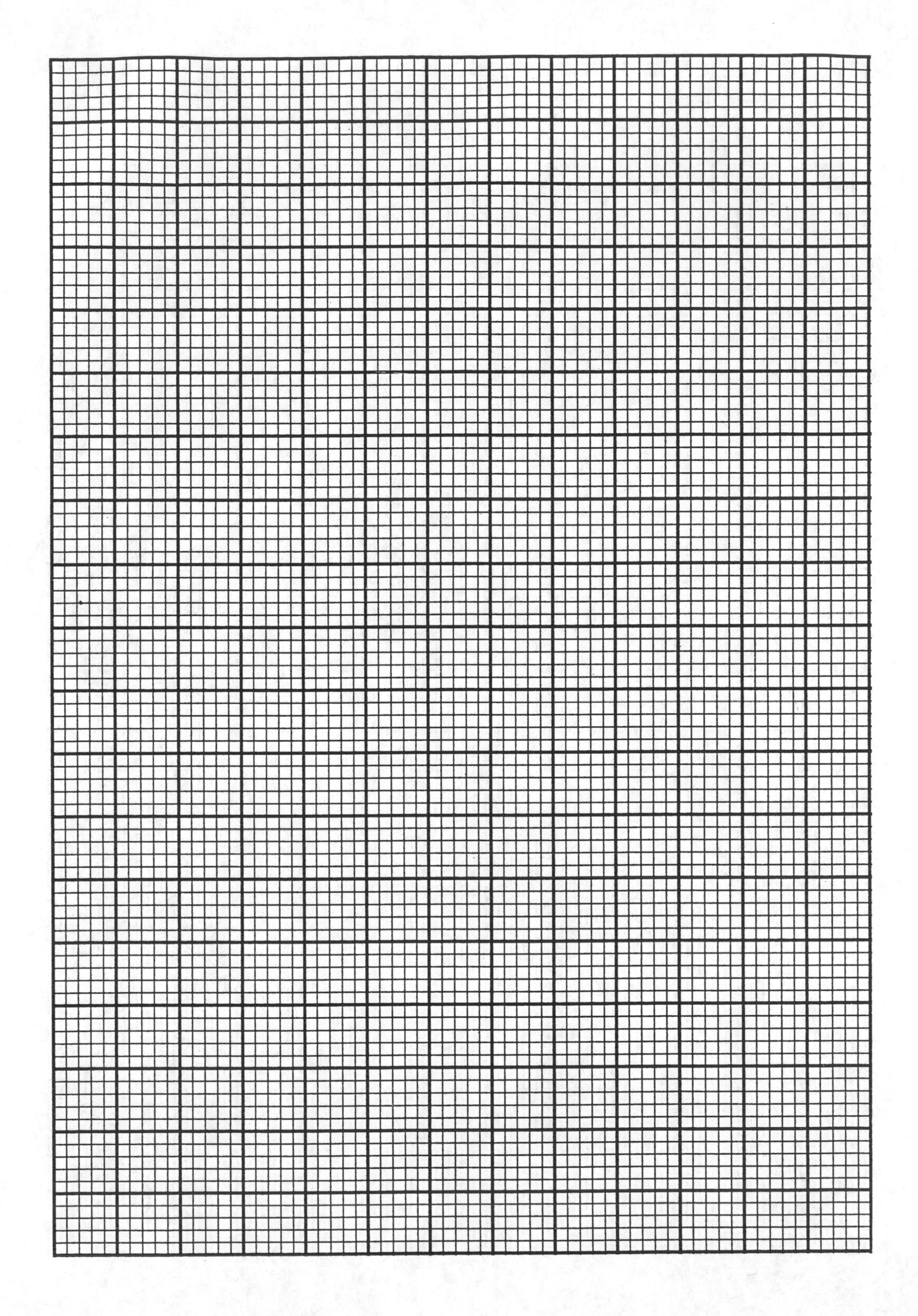

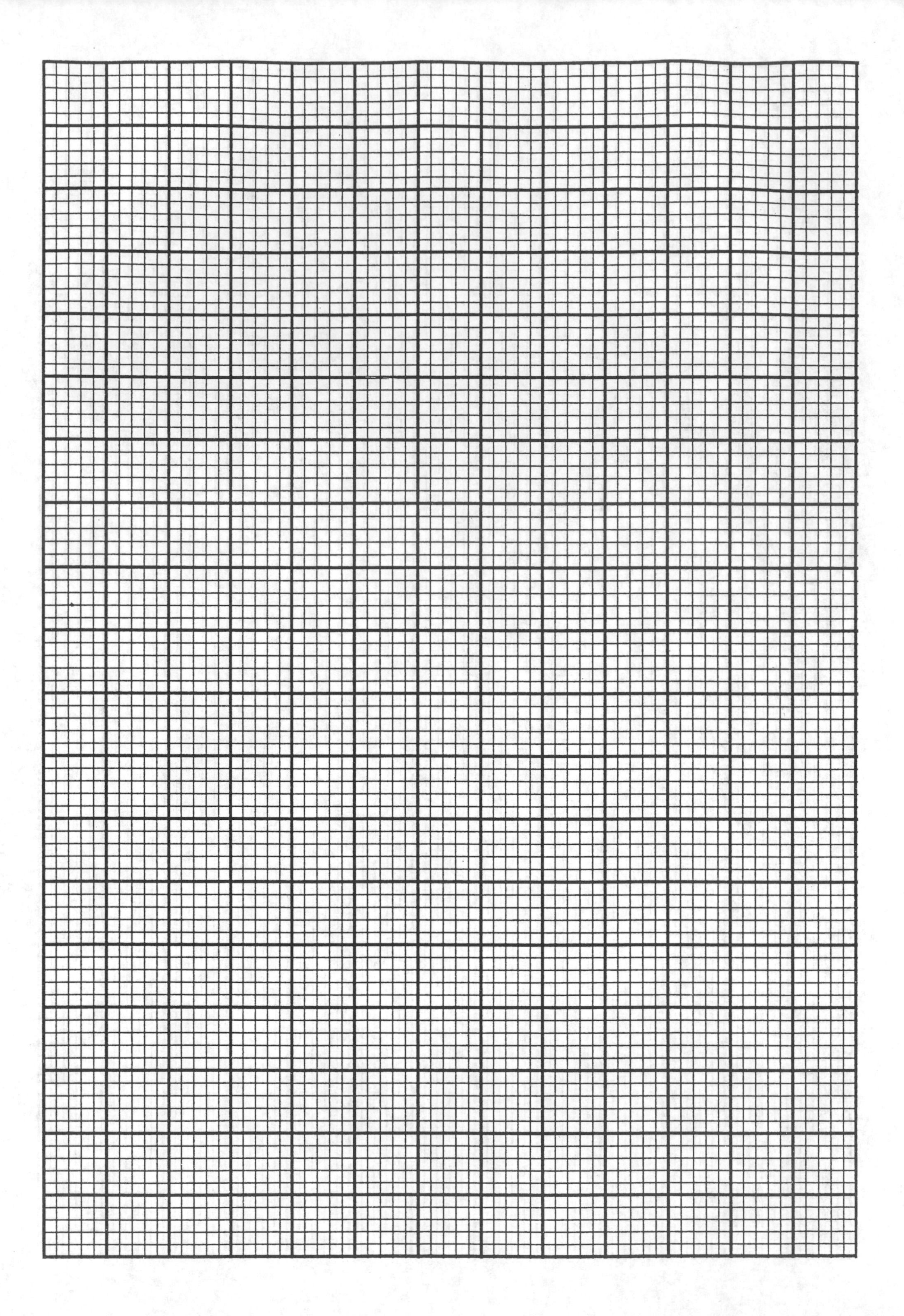

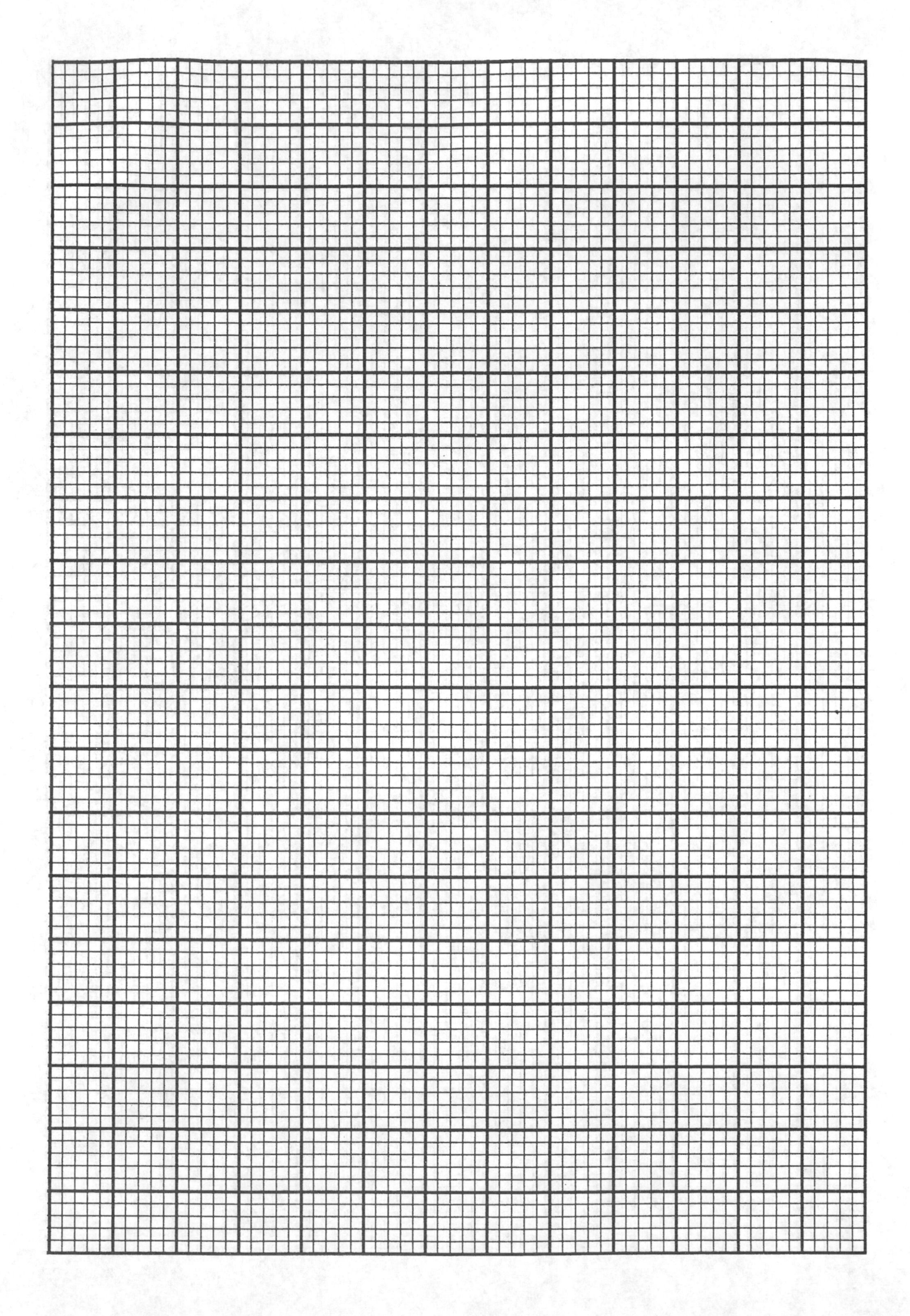

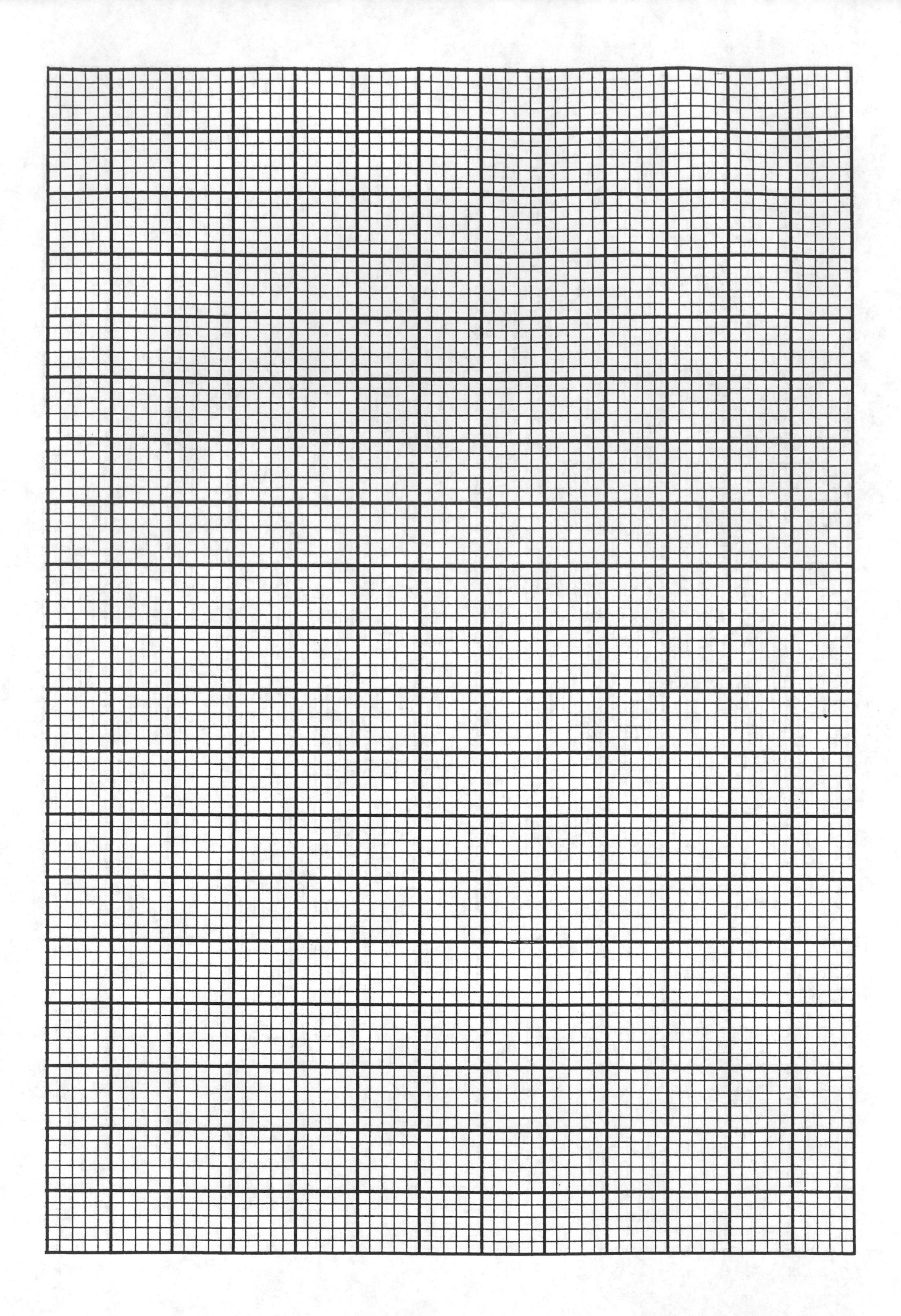

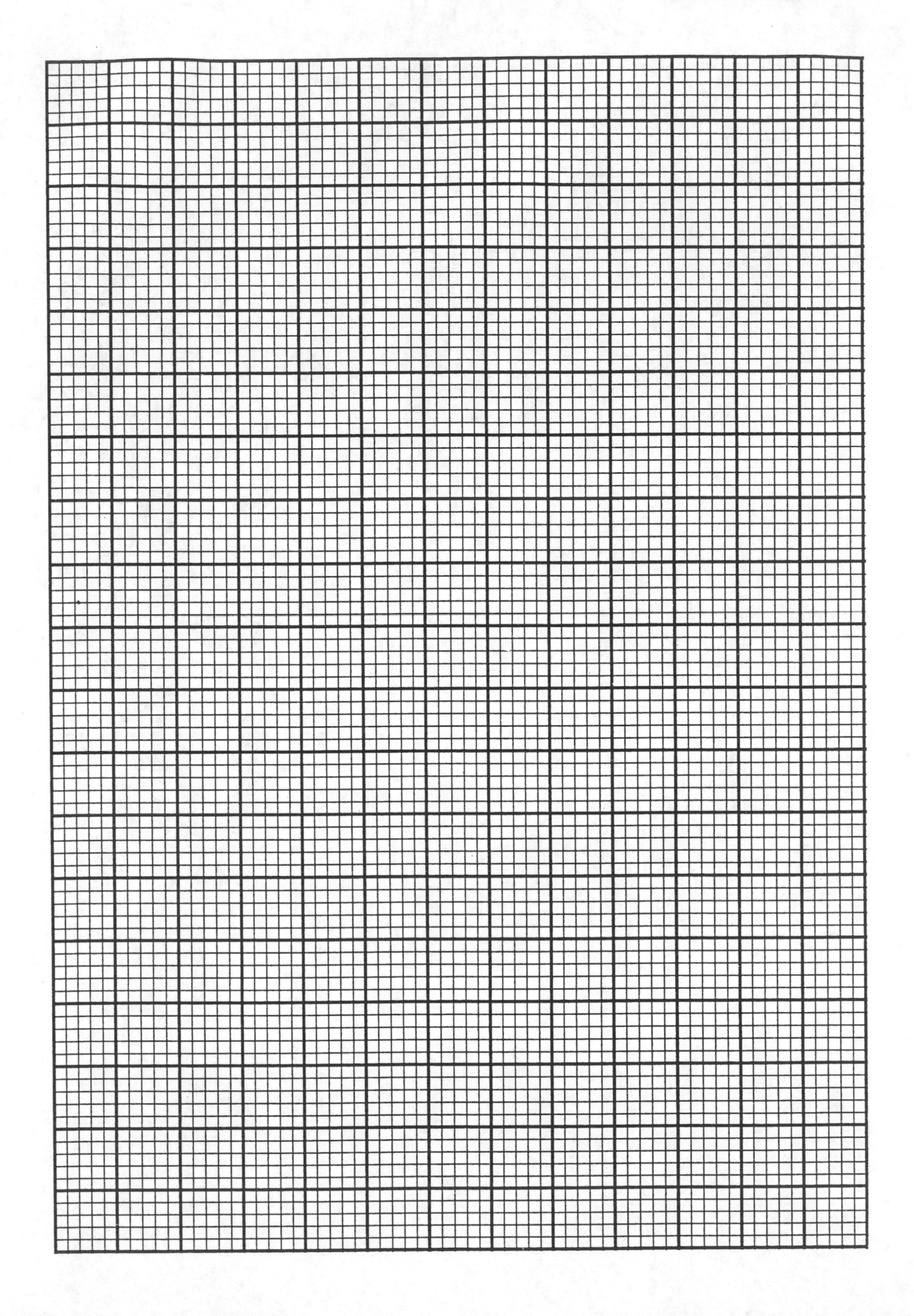

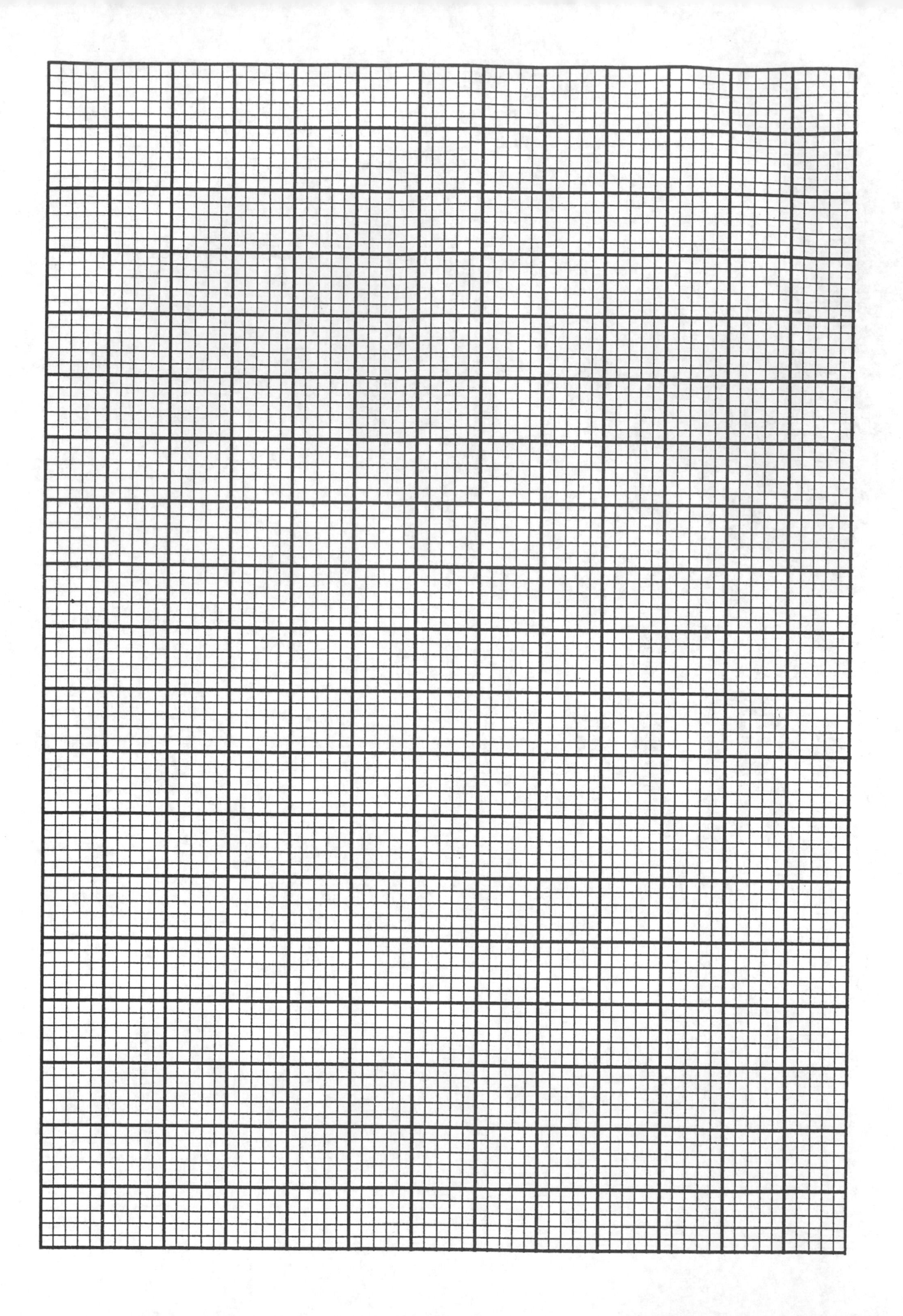

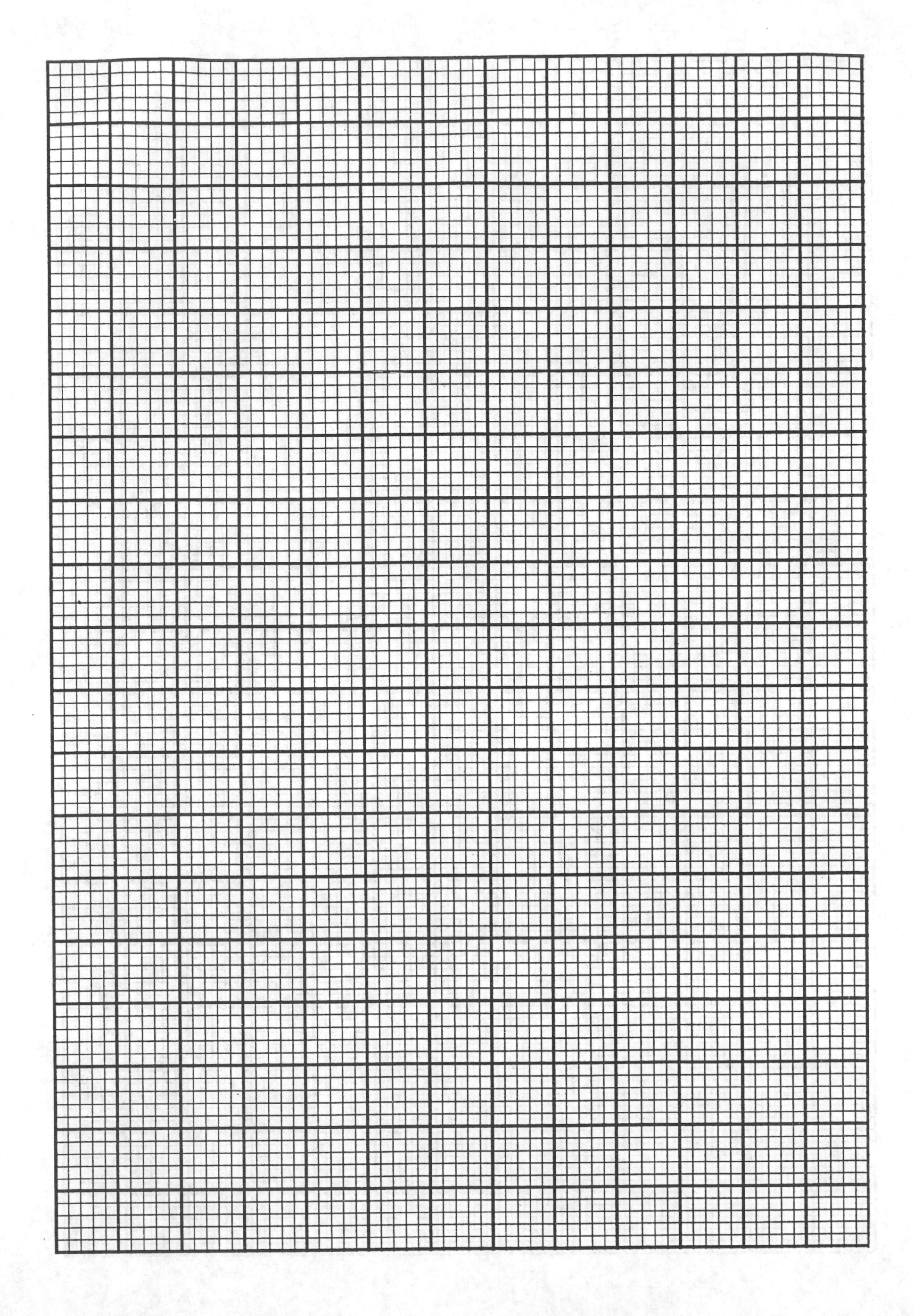

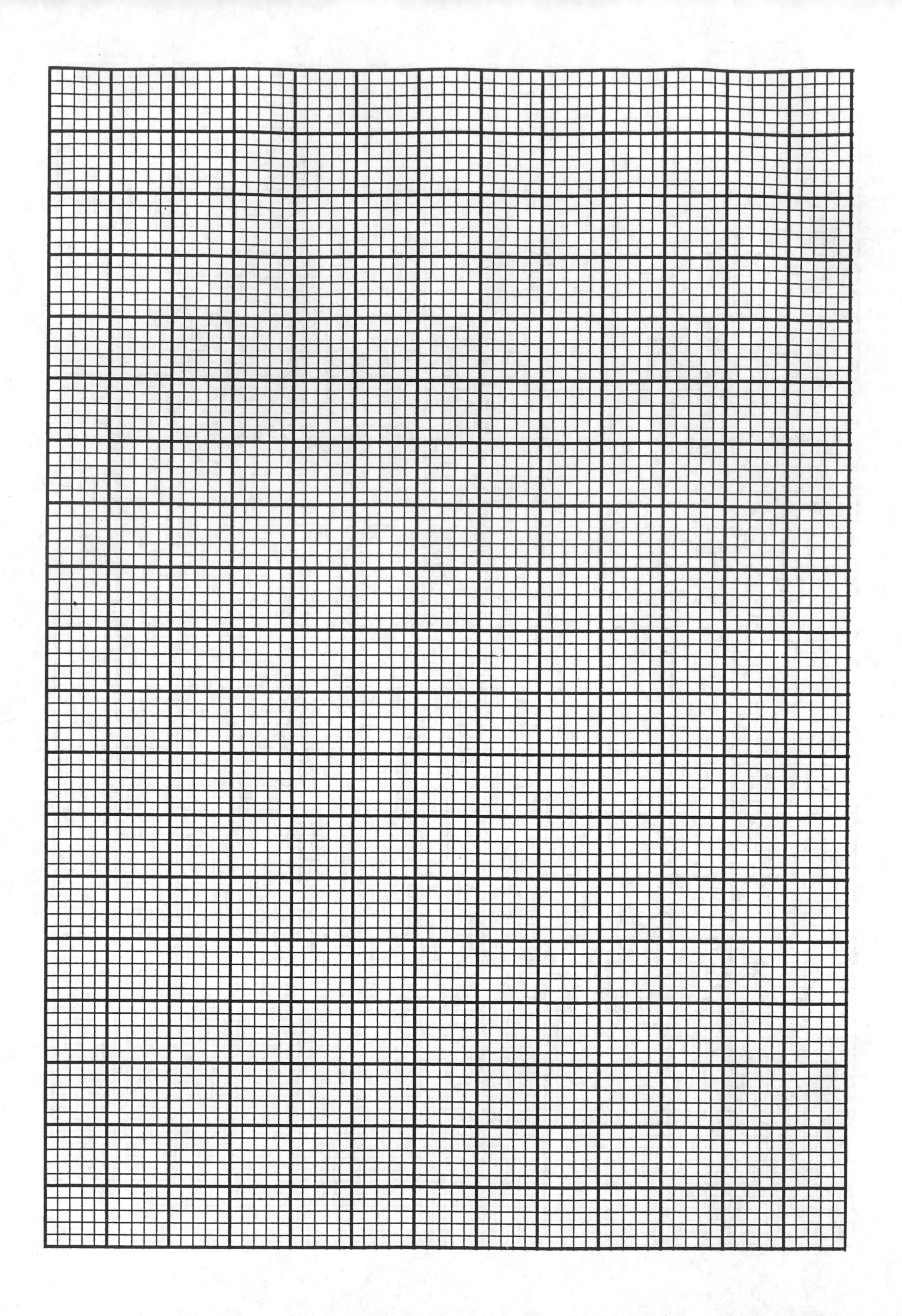

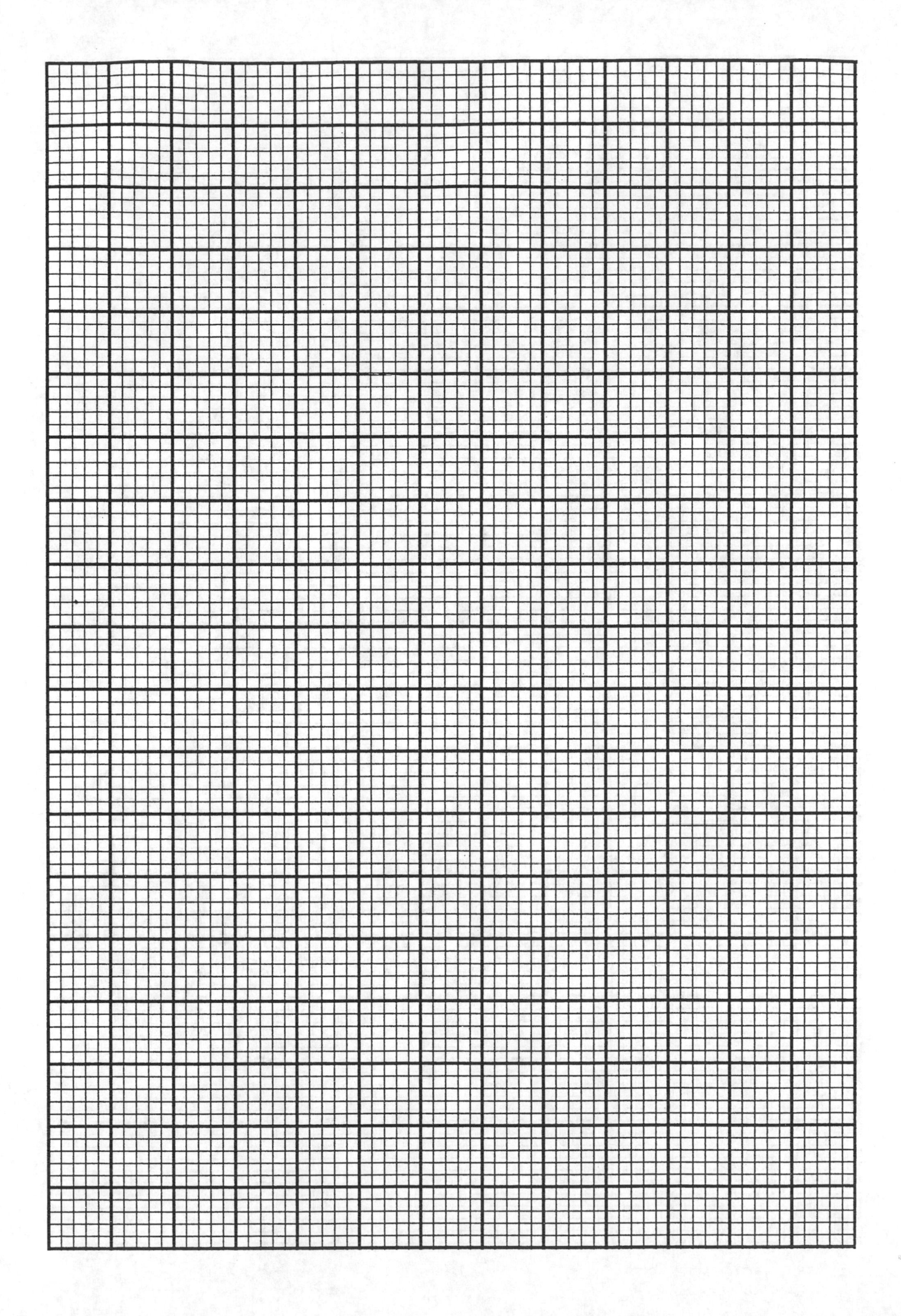

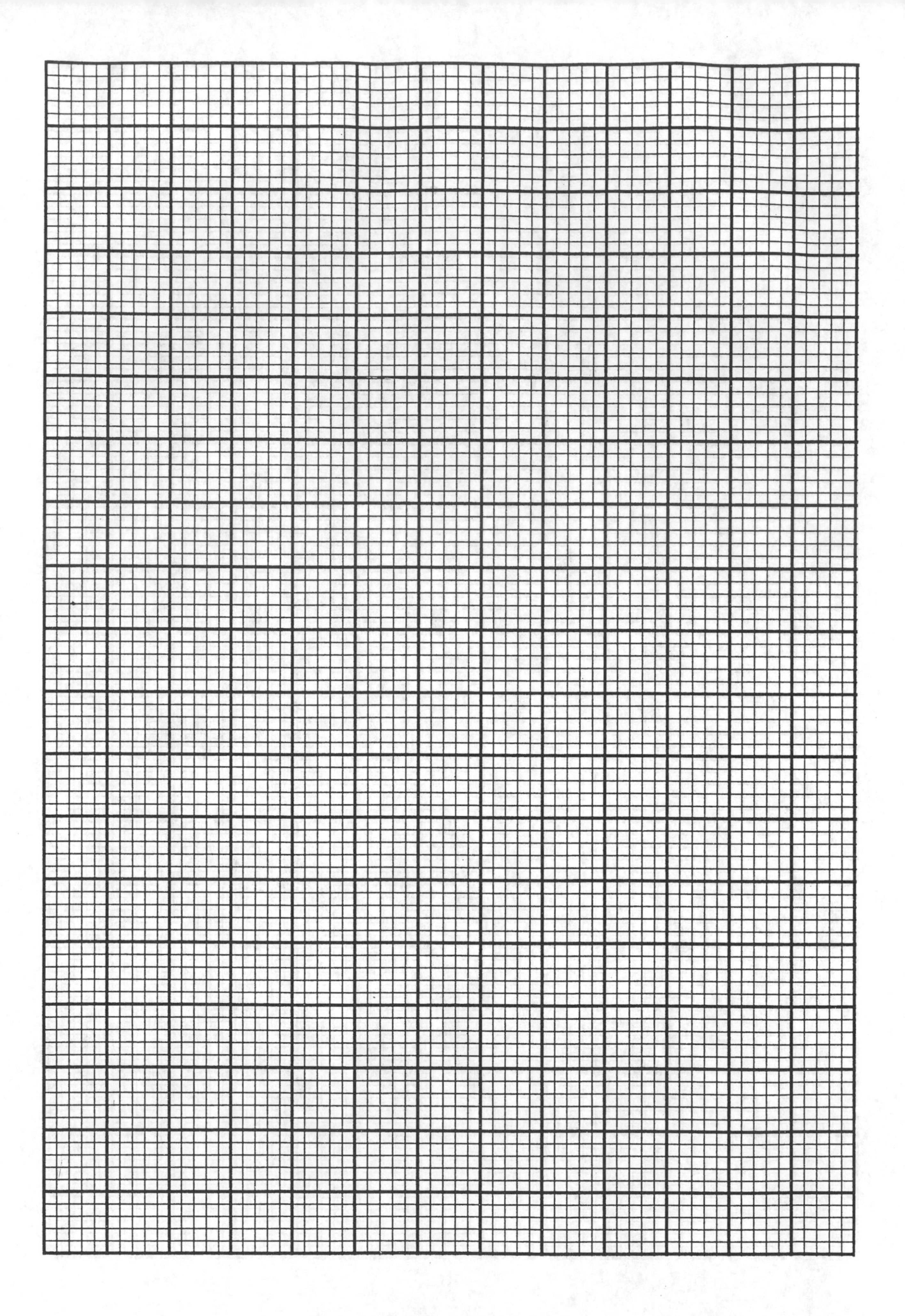

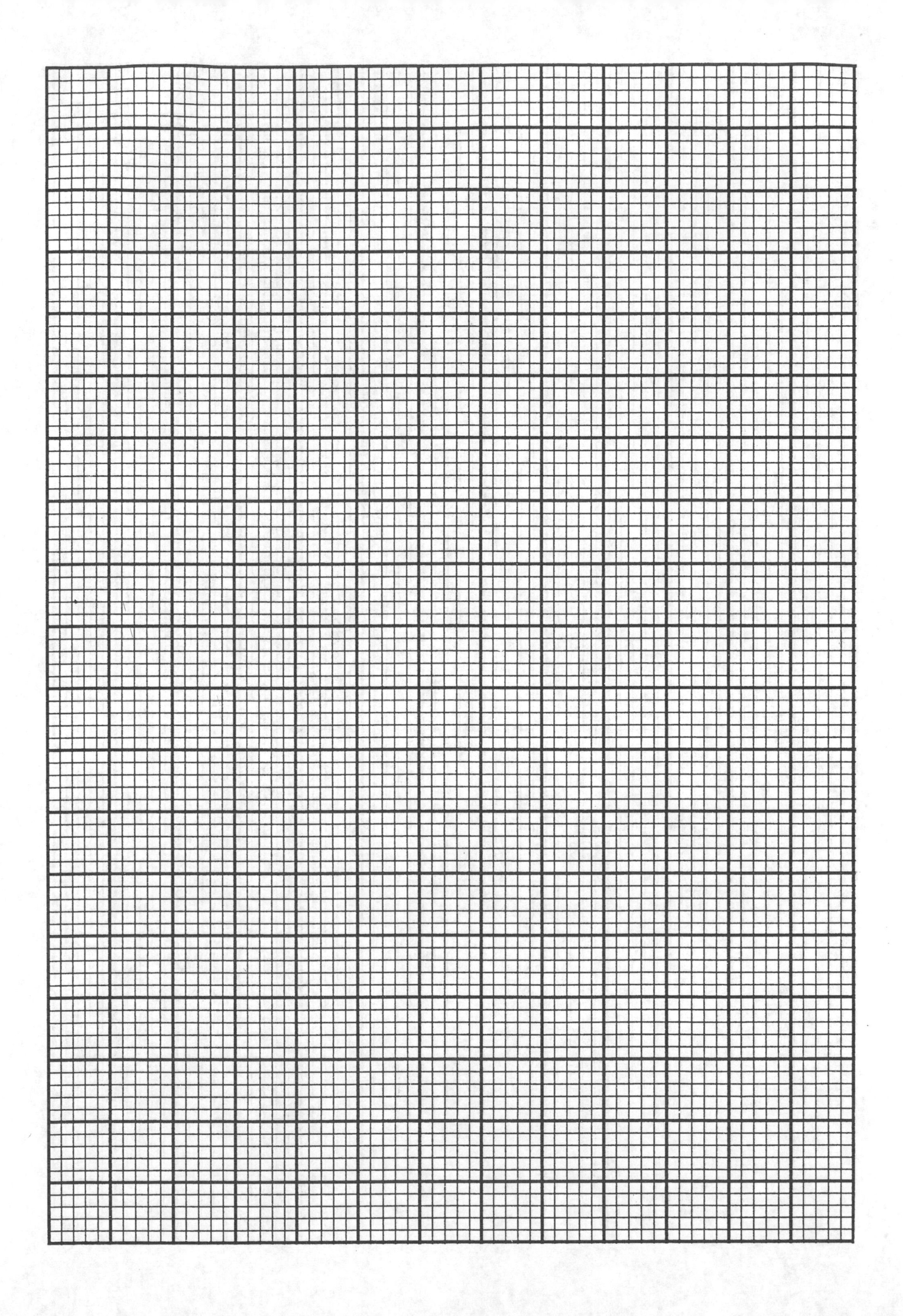

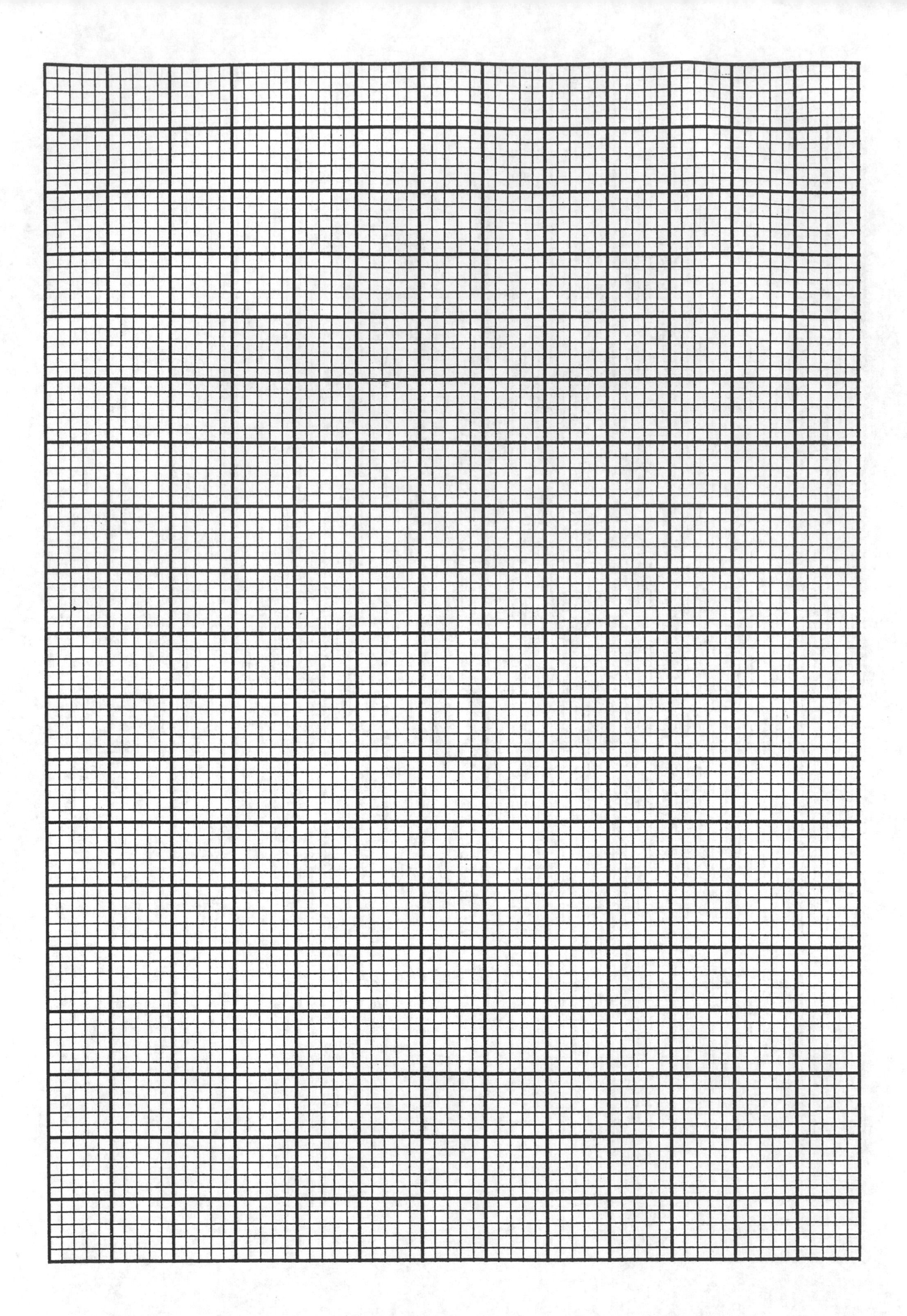

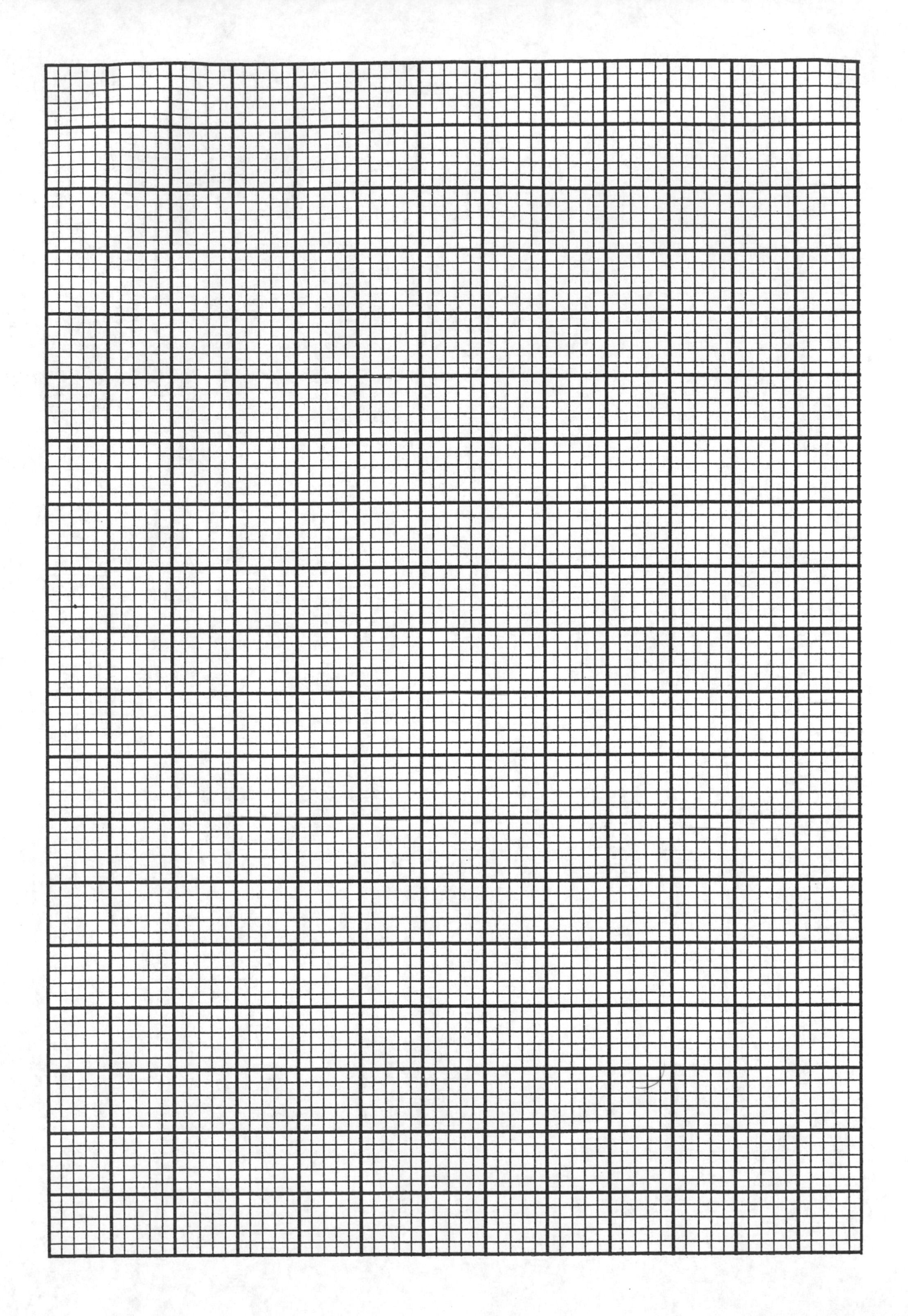

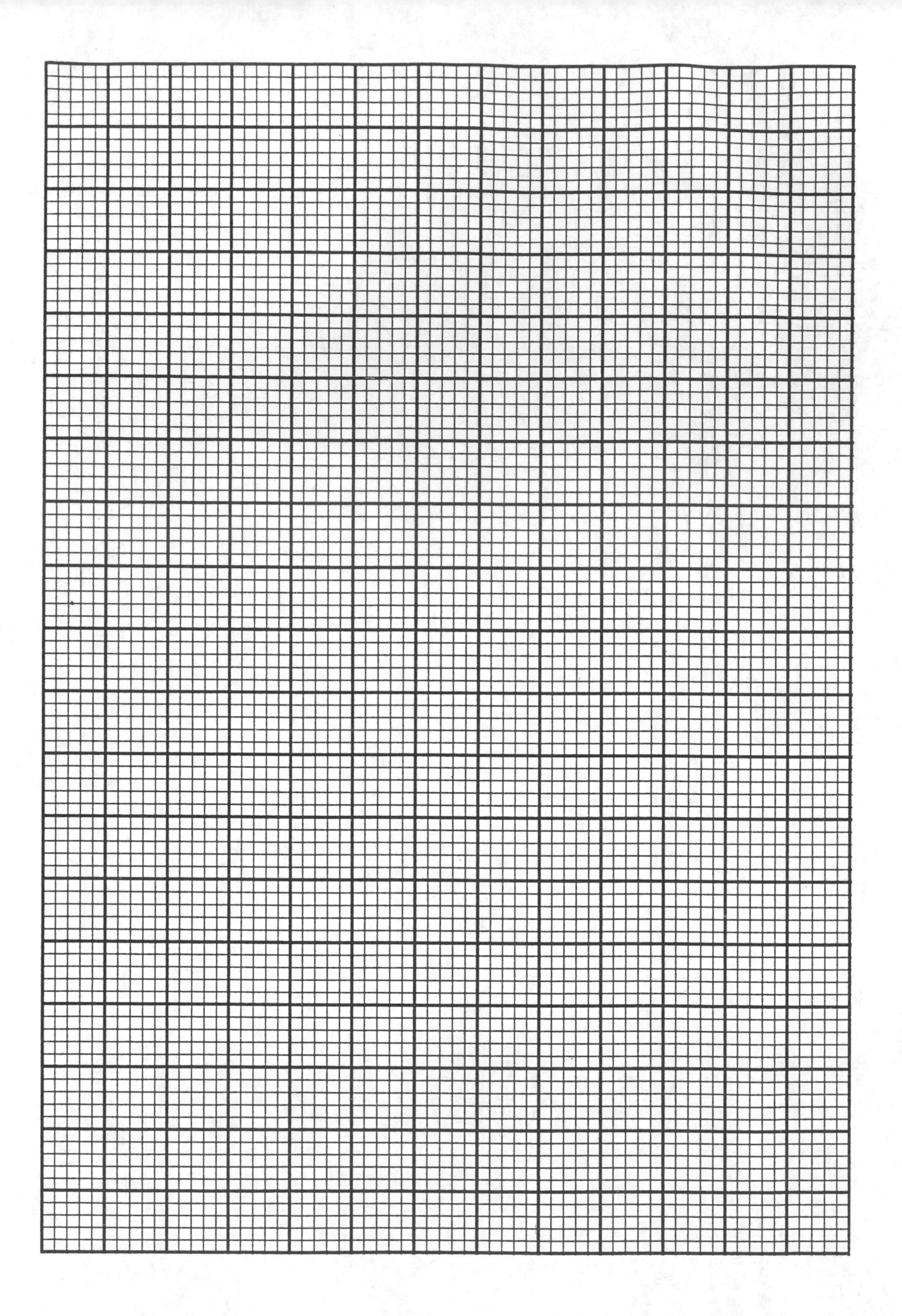

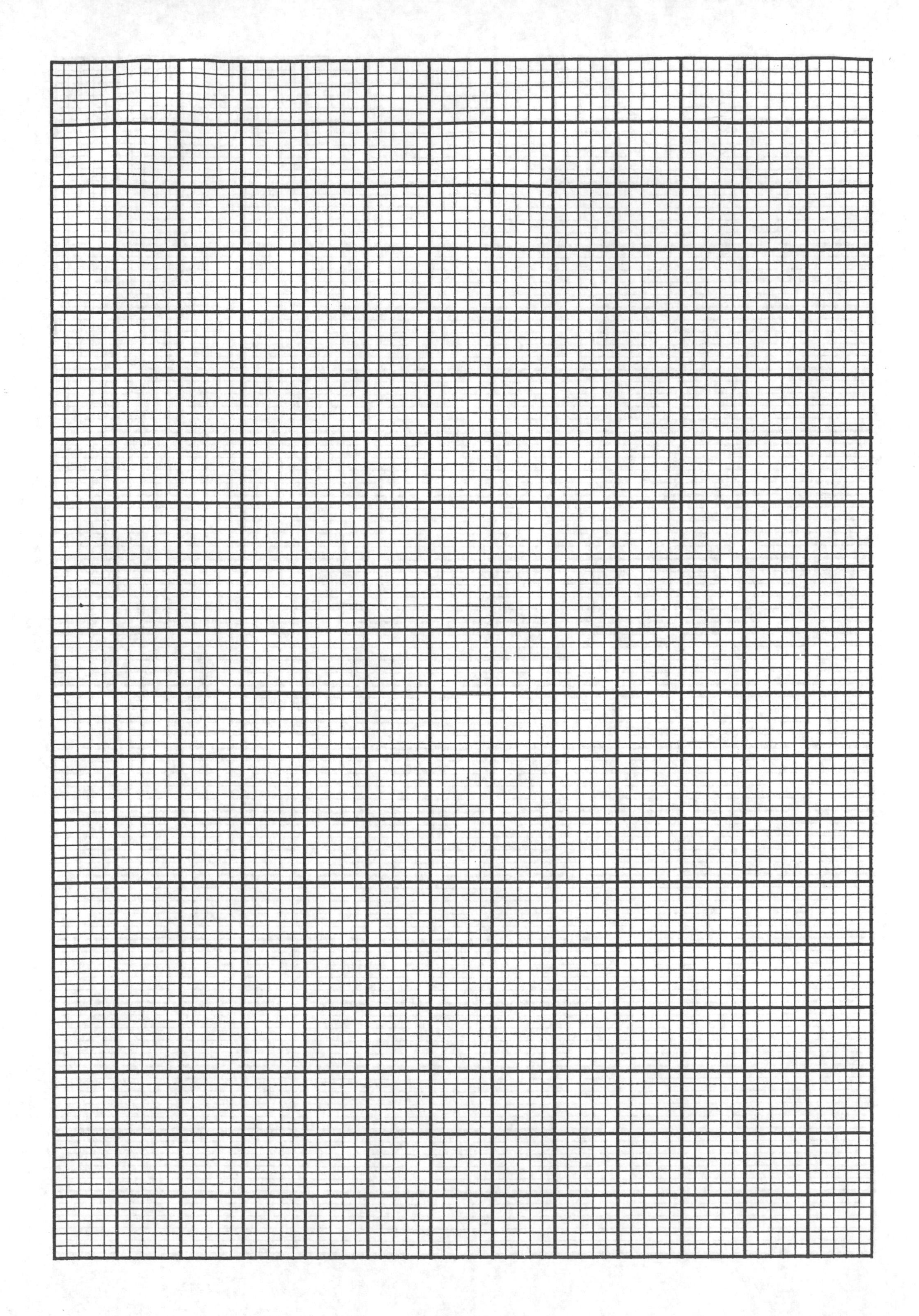

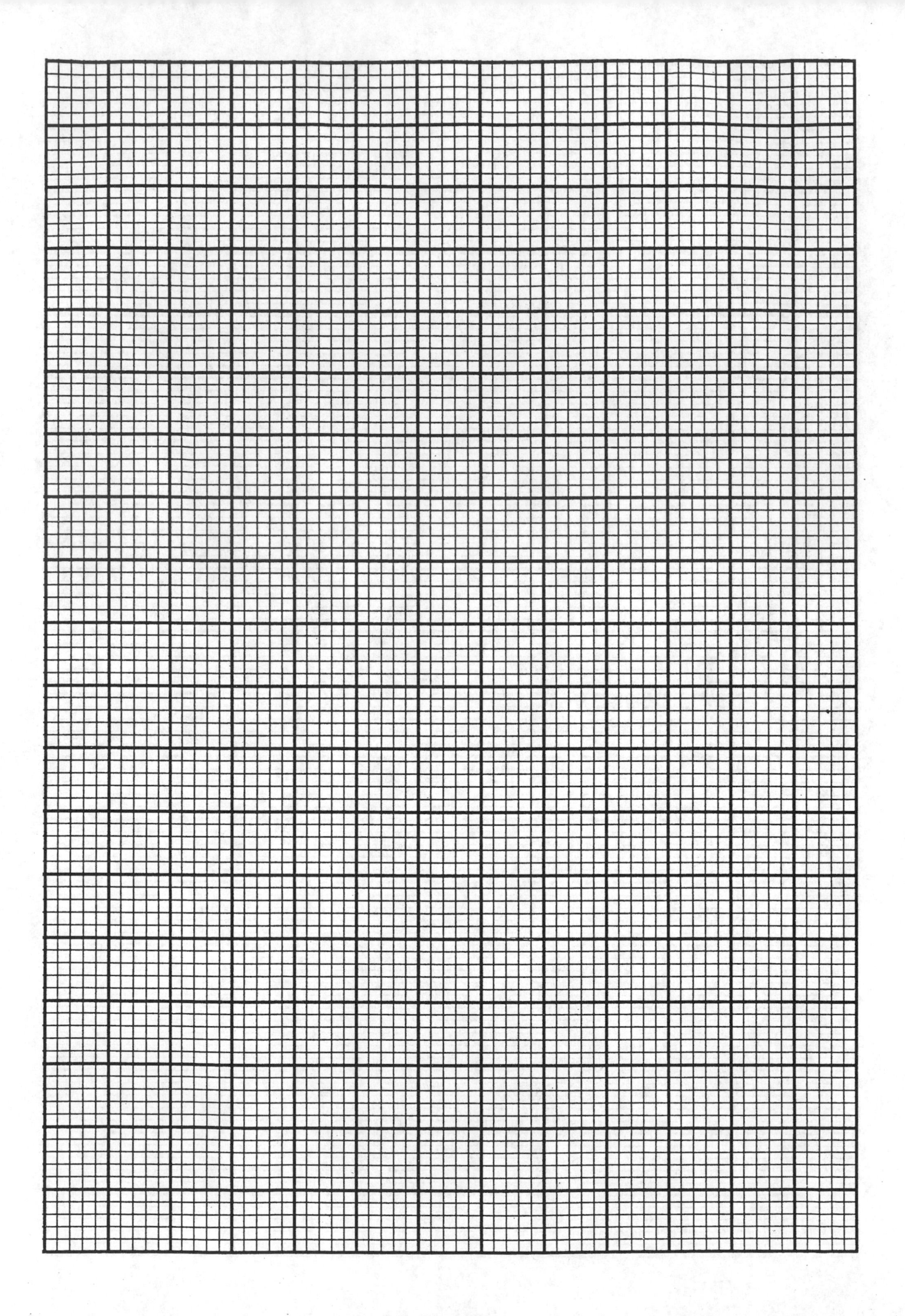